Touobèwèrè Noël SOME

Management of plant formations in the Bontioli wildlife reserves

Touobèwèrè Noël SOME

Management of plant formations in the Bontioli wildlife reserves

ScienciaScripts

Imprint

Cover image: www.ingimage.com

This book is a translation from the original published under ISBN 978-620-3-42402-7.

Publisher:
Sciencia Scripts
is a trademark of
Dodo Books Indian Ocean Ltd. and OmniScriptum S.R.L publishing group

120 High Road, East Finchley, London, N2 9ED, United Kingdom
Str. Armeneasca 28/1, office 1, Chisinau MD-2012, Republic of Moldova, Europe
Managing Directors: Ieva Konstantinova, Victoria Ursu
info@omniscriptum.com

Printed at: see last page
ISBN: 978-620-8-61534-5

DEDICATION

A

My late parents, may the Almighty grant you eternal rest!

My brothers SOME François Bourzié, SOME K. Frédéric, SOME N. Jean Alain, SOME Maxime and SOME Narcisse

My wife TRAORE Fatimata and my daughter SOME Nonfoubome Lydie Merveille.

All environmentalists.

SUMMARY

The Bontioli wildlife reserves, located in south-west Burkina Faso, were set up to protect biological diversity, particularly its wildlife component. The general objective of the present research is to analyze the evolution and management of plant formations in the Bontioli wildlife reserves.

The methodological approach is based on the interpretation of Landsat TM, ETM+ and OLI-TIRS satellite images from 1986, 2002 and 2018, the floristic inventory and socio-economic surveys. The tools consisted of software (ENVI 5.3., ArcGIS 10.3., SphinxV5 and Excel 2013 spreadsheet), questionnaires, interview guides, floristic survey sheets and equipment.

The results show that land-use units underwent significant change between 1986 and 2018. Field and fallow land areas declined by 60.61% and 52.80% respectively in the total and partial reserves. On the other hand, shrub savannah increased by 43.80% and 23.07%, and bare soil by 5.28% and 26.01%. As for wooded savannah, the increase in surface area is perceptible in the total reserve (13.38%). Diversity indices are low according to plant formation. However, gallery forest is the least diverse formation (1.78 bits), while shrub savannah is the most diverse (2.99 bits) on the Shannon scale. Structural parameters indicate stands dominated by young individuals.

All climatic parameters analyzed over the period 1987-2017 show a negative trend, but have had no significant negative impact on plant formations.

In addition, management methods that focus primarily on protecting forest resources and punishing infractions have resulted in low community involvement. This is reflected in the persistence of agricultural clearings in the RPFB. Despite monitoring and awareness-raising efforts by the forestry administration, pastoral practices characterized by selective pruning of plant species fetched by livestock, uncontrolled felling of trees used for fuelwood and timber, uncontrolled gold mining and poaching persist.

In short, the plant formations in Bontioli's wildlife reserves are evolving in step with the state's interventionist policies.

Keywords: protected areas - floristic diversity - anthropization - climate variability - Burkina Faso .

GENERAL INTRODUCTION

Trees and forests are one of mankind's inalienable resources. They are at the heart of animal and plant life, helping to maintain the great balances of land, water and atmosphere, the water and carbon cycle, and the conservation of biological diversity. Vegetation is the foundation of the ecosystem as primary production, on which not only the trophic chain depends, but also the dynamics of biotic communities and their living environments, such as the soil (Cohen, 2008).

By retaining carbon in living biomass, decomposing organic matter and soils, terrestrial ecosystems play an important role in the global carbon cycle. The processes of photosynthesis, respiration, decomposition and combustion maintain the natural circulation of carbon between these ecosystems and the atmosphere (IPCC, 2000).

Forests provide important ecosystem goods to the environment namely soil and water resource conservation, biodiversity conservation, local and even regional climate improvement, carbon storage and carbon cycle regulation as well as energy flow moderation (Aubreville, 1953; Muoghalu, 2014). The economies of most African countries are based on the exploitation of natural resources, particularly vegetation (Ouoba et *al.,* 2014). In addition to these goods, forests and woodland formations provide numerous services to riparian populations in developing countries. Forest resources (woody and non-woody) are used for food, health, housing, tools, improving people's incomes and socio-cultural cohesion (Bordes, 2010; Somé, 2012).

Given the importance of these forest and savannah ecosystems in maintaining biological diversity and their invaluable contribution to human and animal life, their protection is of prime importance.

After the Second World War, the international community turned its attention to the conservation of natural resources as part of a new vision of development. Indeed, it was in 1949 that the notion of sustainable development emerged in the United States of America, with President Harry Truman asserting in his State of the Union address that "all nations, including our own, will benefit greatly from a constructive program for the better utilization of the world's human and natural resources" . [1]

Since then, the sustainable management of natural resources has gradually become a global concern. Voices were raised to denounce unsustainable growth. Clubs, associations and non-governmental organizations sprang up. They have focused their actions on protecting the environment. They include

- of the Club of Rome, created in 1968 to limit the "Trente Glorieuses". In 1972, he published a report entitled "The limits to growth" or "Halte à la croissance?" in which he denounced the impact of uncontrolled industrialization on the environment;

-1972: a United Nations conference on the human environment was held in Stockholm. The first Earth Summit, it laid the foundations for "ecological development";

[1]http://fr.wikiversity.org consulted on 20/7/2016.

- The International Union for Conservation of Nature was founded in 1980;

At African and sub-regional level, organizations such as the Comité Inter-États de Lutte contre la Sécheresse au Sahel (CILSS), of which Burkina Faso is a member, were set up in 1973 following the great drought in the Sahel. They are combining their efforts through ambitious policies such as the Great Green Wall. The overall aim of this gigantic project is to improve the resilience of human and natural systems in Sahelian countries to climate change, through sound ecosystem management and sustainable development of natural resources (water, soil, vegetation, fauna and flora), protection of tangible and intangible rural heritage, creation of rural production and sustainable development hubs, and improvement of living conditions and people's livelihoods (MEDD, 2012).

However, it is clear that these programs and policies are not a panacea for the genuine protection of natural resources, given the forest degradation experienced by African countries. Between 1990 and 2000, the overall reduction in forest cover in West Africa was around 43 million hectares (Pape, 2004; Koffi, 2014). This forest decline, very marked in French-speaking Sub-Saharan African countries, is thought to be the result of the management methods implemented during the colonial period.

In Burkina Faso, the failure of pro-environmental policies and programs and the continuing deterioration of forest resources are said to have prompted the adoption, from the 1980s onwards, of a new vision of forest management based on community participation. This paradigm shift is also rooted in the evolution of natural resource management on a global scale, moving from a conversationalist approach to resources to a community-based and participatory approach, and was rapidly translated into the national field of action (Laurent, 2011).

The Bontioli wildlife reserves, like almost all the country's protected areas, are facing the dual problem of climate change and overexploitation of their resources. These factors affect the spatial occupation and biological diversity of plant formations. It is with this in mind that the present research focuses on the theme: "Dynamics and management of plant formations in Bontioli wildlife reserves (south-west Burkina Faso)".

The study proposes to analyze the evolution and management methods of plant formations in the Bontioli wildlife reserves. To this end, four secondary objectives have been selected: i) - to describe the ecological characteristics and spatio-temporal dynamics of the reserves' plant formations between 1986 and 2018; ii) - to explain the influence of natural factors on the evolution of plant formations; iii) - to examine the impacts of the main socio-economic determinants on the evolution of plant formations; iv) - to analyze the reserve management methods implemented by the riparian population.

The thesis is organized in three parts, subdivided into chapters:

- The first part, subdivided into two chapters, deals with the general framework of the study.
- The second part analyzes the influence of the biophysical and human environment on plant formations in Bontioli wildlife reserves. It consists of three chapters.

- The third part takes stock of management methods and the vulnerability of plant formations in the Bontioli wildlife reserves. It is divided into two chapters.

PART ONE: THE GENERAL FRAMEWORK OF THE STUDY

The aim of this first part is to position this research work both in terms of the schemas that shape its overall reflection and its disciplinary and methodological positioning. It is structured around two chapters:

- ✓ Chapter 1: Theoretical framework and study concepts. In this chapter, the rationale for the study and the research questions were elaborated, and concepts related to the theme were clarified with a view to giving them a precise meaning in the context of this study. Finally, the major biogeographical and demographic theories were recalled and discussed, with a view to understanding the foundations of the evolution of living beings (animals and plants) and the impact of human population growth on natural resources.
- ✓ Chapter 2: Methodological approach. The variables and indicators of the study were defined. On the one hand, spatial and demographic sampling were explained, with answers to the why and how of the choice of survey sites and the actors involved in the research. Data collection and processing tools, equipment and techniques were also explained.

CHAPTER 1: THEORETICAL FRAMEWORK AND STUDY CONCEPTS

This chapter begins by reviewing the current state of knowledge on the subject. It then presents the rationale for the study, from which the research questions, hypotheses and objectives were derived. Finally, key concepts are clarified, and biogeographic and demographic theories related to natural resources are outlined.

1.1. The state of knowledge on the subject

In any scientific work, the author must carefully examine the work already carried out in relation to the theme, in order to give new directions to the study. To this end, this section summarizes the various publications relating to the research theme. It focuses on the following themes.

1.1.1. The state of 2forests worldwide and in Africa

On a global scale, and according to the FAO (2020), forests occupy an area of 4.06 billion hectares, corresponding to 31% of the total land area, and more than half (54%) of these forests are found in just five countries: the Russian Federation (20%), Brazil (12%), Canada (), the USA (8%) and China (5%). The tropical domain has the largest share of the world's forests (), followed by the boreal (), temperate () and subtropical () domains. Some 726 million hectares of forest are in protected areas. Forests have been cut back by 135 million hectares every year since 1980 (Laurent, 2011). At global level, the trend is towards a reduction in the forest area lost, from 5.2 million hectares per year over the period 2000-2010 to 4.7 million hectares over the period

2010-2020. This is due to an increase in forest area and a reduction in the rate of forest exploitation in certain countries around the world.

In Africa, forests and woodlands cover an area of around 675 million hectares, or 23% of the continent and around of the world's forest area. Although according to the indicators, forests have been inexorably shrinking in Africa, net forest loss tended to slow down overall between 1990 and 2010 (FAO, 2011). The breakdown by region is as follows (see Table 1).

Table1 : Forest area in Africa from 1990 to 2010

Sub-region	Area (thousands of ha)			Annual variation (thousands of ha)		Annual variation (thousands of ha)	
	1990	2000	2010	1990-2000	2000 - 2010	1990 - 2000	2000 - 2010
Central Africa	268 214	216 455	254 854	-676	-660	-0,25	-0,26
East Africa	88 865	81 027	73 197	-784	-783	-0,92	-1,01
North Africa	85 123	79 224	78 814	-590	-41	-0,72	-0,05
Southern Africa	215 447	204 879	194 320	-1 057	-1 056	-0,50	-0,53
West Africa	91 589	81 979	73 234	-961	-875	-1,10	-1,12
Total Africa	749 238	708 564	674 419	-4 067	-3 414	-0,56	-0,49
World	4 168 399	4 085 063	4 032 905	-8 334	-5 216	-0,20	-0,13

Source: FAO, 2011

Forest formations regressed significantly in all African regions over the two decades considered in the table. Between 2010 and 2020, Africa lost 3.9 million hectares, followed by South America 2.6 million hectares (FAO, 2020). By contrast, forest cover is increasing in North Africa and stable in South Africa (Madon, 2017). The degradation of tropical forests (wet and dry) is undoubtedly one of the main causes of biodiversity erosion. The decline in forest area, particularly in Africa, is multifactorial. Nevertheless, the need for agricultural land and the dependence on wood energy of the population of sub-Saharan Africa, which is essentially rural (62% in 2015) (FAO, 2018), seem to be the main causes. However, the question of political governance seems to cut across all development issues in African countries. It also affects the forestry sector.

1.1.2. The creation and evolution of protected areas worldwide

The notion of protecting specific areas for targeted exploitation has always been the prerogative of all human societies. However, the first laws to protect forests are thought to have been enacted in Mesopotamia, around 2,700 BC, although other examples exist in Egypt and India (Doumenge, 2021). Since the creation of the first legally-recognized national park, Yellowstone Park, in 1872 by the US Congress, mankind has become progressively aware of the need to protect nature, as the establishment of protected areas intensified over the course of the 20th century. These nature reserves were mainly found in America and Africa. It wasn't until the 1960s that other continents stepped up the pace of protected area creation.

The first national park created in Africa, Kruger Park, dates back to 1898 and is located in South Africa. It was followed by Albert National Park, created in 1925 in the Belgian Congo (now the Democratic Republic of Congo) under colonial rule (IUCN, 1999). In French-speaking Africa, there are three generations of protected areas: between 1930 and 1950 during the colonial era, from 1960 to 1990 after independence and 1990 to 2001 after the Rio conference. In English-speaking Africa, on the other hand, the majority of protected areas were created between 1950 and 1970 (Mengue-Medou, 2002). There are over 7,000 protected areas in Africa, to which must be added areas that contribute to conservation and are mainly managed for economic purposes, such as hunting zones, classified forests and natural resource management zones (Chardonnet, 2019).

The creation of protected areas on the Asian continent began during the colonial period. The Banyuwangi game reserve in Java, created in 1919, and the Krau wildlife reserve on the Malay peninsula, created in 1923 (Déry, 2008). However, it was only after the 1960s and precisely at the beginning of the 1990s, in the wake of the 1992 Rio Conference, that the creation of protected areas intensified in Southeast Asia.

1.1.3. The state of the world's protected areas

Europe, home to 43,000 protected areas, ranks first in terms of the number of listed protected areas. In second place is Eurasia with 18,000, followed by North America (13,000), Australia (almost 9,000), Southern and Eastern Africa (4,390), Central and West Africa (2,600) and the Pacific, which has the fewest protected areas (320). However, Central and South America have the largest protected areas, with of the surface area of each of these regions. They cover 1.6 million km², or 14.of southern and eastern Africa (Martinez & Triplet, 2009). These protected areas are divided according to the type of environment in which they are found. Table 2 shows the main types of environment and the surface areas represented in protected areas.

Table2 : Main types of environment and surface areas represented in protected areas

Habitat types	Total area (km^2)	Protected area (km^2)	Percentage protected
Tropical rainforest	10 390 000	2 471 000	23,8
Tropical dry forest	2 716 000	339 000	14,7
Savannah	15 368 000	1 878 000	12,2
Scrub	5 611 000	692 000	12,3
Wetlands	3 429 000	434 000	12,7
Desert	45 474 000	4 589 000	10,1

Source: Adapted from Mulongoy & Chape, 2004)

Today, there are over 265,000 established or proposed protected areas worldwide, covering 15.7% of land and 7.of sea. Currently, over 22.5 million km^2 on land and 28.1 million km^2 in marine and coastal environments are protected (Doumenge, 2021).

1.1.4. Forest governance

The management of protected areas is governed by social, cultural, economic, legal and environmental principles, insofar as trees contribute to human and animal life. Legislative provisions have therefore been drawn up for the efficient management of forest resources.

1.1.4.1. The evolution of forestry policies and legislation in French-speaking Africa from the colonial period to the present day

Forestry legislation in French West Africa (AOF) has its roots in French forestry policies. A so-called "conservationist" movement emerged in France in the ^{16th} century and imposed its vision of forest protection in its colonies until 1980; the idea being to make the natural capital of the forest, in this case, impervious to human intrusion (Laurent, 2011; Koffi, 2014).

1.1.4.1.1. Tropical forest policies and legislation in French-speaking Africa during the colonial period

According to Madec (1997), tropical forestry regulations originated with a decision by the Governor of Cochinchina on May 18, 1862, when the territory was still only partially conquered. However, in the almost total absence of indigenous regulations, it was the metropolitan forestry code of May 21 1827 that was to be the essential source of tropical forestry legislation. The decree of August 5, 1900 on the Senegalese forestry regime and dependencies was extended to vast regions, including Sudan (Mali), Guinea and Upper Volta (Burkina Faso). According to Kiéma (2007), this decree was revised on March 24, 1901, and again on July 18, 1912. This Code created permits, concessions, protected areas and zones, as well as usufructuary rights (Ribot, 2001).

In reality, these decrees were simply provisions contained in the Code de l'Indigénat, from which they derived. This Code had been promulgated in 1887, then revised and corrected in 1905, 1912, 1921 and 1924. In the end, it was the decree of July 4, 1935 that provided the legal basis for forestry action in Burkina Faso, as in the rest of the French West Indies.

The main argument underpinning the need for forestry legislation to benefit tropical forests was later reiterated in the decree of May 20 1955 in the following terms: "The protection rules envisaged tend to safeguard the general interest, having regard to the scientifically recognized influence of afforestation on the protection of soils against erosion, on the maintenance of springs and rivers, as well as of the general characteristics of the climate, particularly with regard to the rainfall regime" (Guillard, 2014).

1.1.4.1.2. Tropical forestry policies and legislation in French-speaking Africa in the post-colonial period

After the colonial period, legislation in most French-speaking African countries was inspired by international policies on the conservation and management of natural resources. The ecological, socio-economic, cultural and scientific functions of forests now take pride of place.

In the recent past, most African countries have reformed their forestry legislation to varying degrees. The major trends emerging from the reformed African forestry legislation are essentially threefold: (i) forest management based on the principles of sustainability, with more systematic use of management plans; (ii) forest management with an increasingly local and private dimension; (iii) greater recognition of the environmental functions of forests (Textile & Kante, 2005).

From 1980 onwards, a new paradigm emerged in forest resource management. This involved a shift from an exclusive protectionist approach to a participatory, community-based approach now advocated by the international bodies responsible for monitoring and supporting protected areas (Giraut & *al.,* 2004). The participatory approach to natural resource management makes it possible to take account of the local population's aspirations in terms of development, and the associated rights and obligations.

1.1.4.2. Evolution of the political and legislative framework for forest management in Burkina Faso: from the pre-colonial period to the present day

Environmental law in Burkina Faso is closely linked to the colonial and post-colonial periods. However, it should be emphasized that in traditional societies - the pre-colonial period - the premises of environmental law prevailed, notably with the regulation of access to and exploitation of natural resources, the collective appropriation of natural resources, and the institution of totem poles. The strict and rigorous regulation of fishing and hunting, the establishment of sacred forests and ponds, and the creation of totem poles were all forms of normative regulation concerned with preserving the environment. With colonial penetration, traditional rules disappeared and were replaced by colonial regulations (Djiga, 2012).

Forests in Burkina Faso are highly dynamic open spaces, evolving at the pace of natural and man-made disturbances. It is therefore with a view to the proper management of these environments that, since the colonial period, the colonial authorities have taken the initiative of creating protected areas. As early as 1926, the first protected areas were created under the name of "refuge parks" (Fournier & Sinsin, 2007; Tianhoun, 2012). It should be remembered, however, with Giazzi & Tchamie (2007), that the legal concept of protected areas was introduced in French West Africa by the decree of March 10, 1925, which instituted refuge parks. However, the adoption of the Convention for the Protection of Fauna and Flora in Africa in London in 1933, to "promote" an ideology common to all the colonies, was a decisive turning point in the creation of protected areas in the French colonies

(Diallo, 2011). Indeed, the forest estate in its modern acceptation was born with the constitution of a forest estate in 1933 (Sawadogo, 2006).

It should be noted that the country had no forestry legislation of its own, and was part of the natural resource protectionist paradigm derived from French forestry law. The institutional architecture of forest governance in French-speaking African countries is based on a common colonial heritage, that of 18th-century French colbertism, founded on a forest police force responsible for guaranteeing the physical integrity of the forest against any use by rural populations (Koffi, 2014).

Three major periods have marked the evolution of forestry policy in Burkina Faso:

✓ In colonial times

It was the decree of July 4, 1935 that paved the way for vigorous intervention by the colonial authorities, and led to the creation of the large classified estates. The primary reason for classifying forests as state-owned was to reduce deforestation. The main objectives of these classifications were :

- the installation of climatic plant barriers to reduce the influence of drying winds from the north (harmattan);

- protection of major watercourses ;

- building up wood reserves for charcoal production to supply the steam trains of the Régie Abidjan-Niger (RAN) and the planned railroad line linking Bobo-Dioulasso to Ségou (Republic of Mali);

- the creation of wildlife reserves for hunting;

- production of timber and service wood for local government development needs (MECV, 2007a).

The creation of protected areas therefore intensified in Upper Volta, now Burkina Faso, until the end of the 1950s.

✓ After

The protectionist policy introduced by the colonial rulers continued until the early 1970s, when a severe drought in the Sahel led to the adoption of a more active policy for the development of forest resources: this was the era of large-scale plantations "to satisfy the need for firewood and utility wood and to combat desertification". The failure of this strict state interventionist policy gave rise to the community forestry approach, with the flagship "Village Woods" program launched in 1978:

- a vast eucalyptus reforestation campaign;

- the promotion of improved fireplaces to reduce household consumption of firewood;

- the creation of the Centre National de Semence Forestière (CNSF).

The village wood project did not really see the light of day until 1984. The main guidelines emphasized the need for an approach to forestry interventions involving the conscious participation of rural populations in the identification and implementation of actions (Semeni, 2007).

Nevertheless, the forest protection strategy implemented was essentially based on the repression of infringements of the law and the use of "mise en défens" (Bakouan, 1995). The "mise en défens" technique consists in restoring ecosystems by more or less totally prohibiting human exploitation of resources in a territory (Somé & *al.*, 2011). Not only has this approach not led to efficient forest management, but worse still, it has contributed to developing a negative perception of the forestry service among the population. What's more, the repression-based approach has never been able to combine the exploitation of the resource with its conservation (Diarra, 1999). During the first decades after Burkina Faso's independence, forest management was not a national priority, as it was only in 1976 that the Ministry of the Environment and Tourism was created (Ouédraogo, 2001).

- ✓ The participatory approach was tested from 1986 onwards.

This latter approach is now the main strategic thrust of national policy on natural resource development (Gautier & Compaoré, 2006). Overall, it can be understood, according to Zoungrana (1995), as confirmation of the leading role of villagers in realizing their individual and collective aspirations, with or without external support. It lays the foundations for endogenous, integrated and sustainable development, tailored to the aspirations of the people. Participatory management refers to the processes and mechanisms that enable people directly affected by the use of forest resources to be involved in decision-making on all aspects of forest management, from resource management to the formulation and implementation of institutional frameworks. [2]

According to Ouédraogo (2004), the most important events in the evolution of forest management in Burkina Faso are as follows:

[2] http://www.fao.org/forestry/participatory/fr/ consulted on 10/9/2016.

1981: development and publication of a national policy for the management of classified forests, the main objective of which was to satisfy the population's need for timber forest products, while preserving the environment;

1982: adoption, by the Comité de la Recherche Forestière, of research protocols relating to vegetation dynamics (evolution of coppice, determination of favourable periods for cutting wood in relation to regeneration, effects of fire on vegetation);

1983: setting up of experimental vegetation dynamics systems in forests located in different ecological zones of the country, and above all the elaboration of a technical document entitled "land evaluation methodology applied to forest management in Burkina Faso", which is a systematic analysis procedure for socio-economic data;

1985: on April 22, three political measures were taken to protect the environment, known as the "Three Struggles"

- Animal control,
- Bushfire control,
- Combating illegal logging.

1986: start of the first participatory forest management activities with the "development and exploitation of forests to supply Ouagadougou with firewood" project (UNDP / FAO / BKF / 85 / 011);

1989: development of a national forest management policy, which represented an extension of the national classified forest management policy, geared towards defining the role of forestry in land use and management;

1995: national forestry policy drawn up;

1996: national forest management program drawn up;

1997: adoption of law n°006/97/ADP of January 31, 1997 on the forestry code

Burkina Faso;

1998: adoption of decentralization guidelines;

2001: drafting of the Decentralized Rural Development Policy Letter ;

2002: development of a methodological guide to forest management in Burkina Faso.

These policies have been readapted to take account of the international context and the imperatives of rural populations' needs. New legislation has been passed, including Law N°0034-2009/AN on rural land tenure in Burkina Faso, which clearly defines the principles and rules governing land tenure in Burkina Faso, with a view to securing it. The 1997 forestry code was revised and adopted in 2011 to ensure more efficient management of forest resources.

All the laws and regulations governing environmental management in Burkina Faso, and in particular forest management, are based on Article 29 of the Constitution of Burkina Faso of June 11, 1991, which : "The right to a healthy environment is recognized; the protection, defense and promotion of the environment are the duty of all".

1.1.4.3. International conventions ratified by Burkina Faso

The international community has long been aware of the need to protect the environment by creating and restoring protected areas. It should be noted that it was as early as the 13th century, in Europe, that the risk of a shortage led to the appearance of the first regulations aimed at perpetuating forest resources, in this case the French Royal Order of 1219 (Puyo, 2001). However, it was in the United States of America that the 888,000-hectare Yellowstone Park was created on March 1, 1872 by the US Congress (Raffin, 2005). It was the world's first legally instituted park. Over the years, this desire has been clearly reflected on the international scene, with the creation of numerous nature reserves in America, Africa, Asia and Europe. This surge in the number of nature parks is associated with the development and adoption of international legal instruments for the management of forest resources. Notwithstanding their non-binding nature, these instruments provide a reference base for the planning, implementation and protection of forests by the various States.

The first United Nations Conference on the Human Environment, which became the first Earth Summit, was held in Stockholm in 1972.

Some twenty years after Stockholm, the United Nations Conference on Environment and Development (UNCED, generally known as the Earth Summit), meeting in Rio de Janeiro in June 1992, adopted two important non-binding instruments: the Rio Declaration and Agenda 21, also known as the Earth Action Plan.

In addition, UNCED developed a statement of forest principles that remains relevant as a non-binding declaratory instrument for forest conservation. This statement, formally entitled "Non-legally binding but authoritative statement of principles for a global consensus on management, conservation and sustainable development" (UN, 1992, Annex 3), represented an important milestone in the consideration of conservation and sustainable development in the forest sector. It was not legally binding because efforts to adopt a treaty on forest conservation were not sufficiently advanced. Of the Declaration's fifteen (15) principles, two (2) are particularly relevant to forest protected area policy and law:

- Forest management should be integrated with the management of contiguous areas to preserve ecological balance and sustainable productivity.

- National policies and legislation aimed at the management, conservation and sustainable use of forests should include the protection of representative or ecologically viable forest types, including virgin or old-growth forests and other forests of cultural, spiritual, historical, religious or other value, of national importance (UN, 1992, Annex 3, principle 8(e), 8(f)) (Barbara, 2012).

Adopted at the Earth Summit in Rio de Janeiro in 1992, and entered into force on December 29, 1993, the Convention on Biological Diversity (CBD) is a legally binding international treaty. In the preamble, the signatory countries affirm the urgent need for all humankind to conserve biological diversity, given its importance for the evolution and protection of the biosphere.

It has three (3) main objectives: the conservation of biological diversity; the sustainable use of biological diversity; and the fair and equitable sharing of the benefits arising out of the utilization of genetic resources. Its overall aim is to encourage measures that will lead to a sustainable future. The conservation of biological diversity is a common concern of mankind. The Convention on Biological Diversity covers all levels of biological diversity: ecosystems, species and genetic resources .[3]

Having ratified the Convention on June 12, 1992, Burkina Faso has developed strategies for its operationalization. Two (2) flagship documents have been drawn up for this purpose, including Burkina Faso's National Strategy and Action Plan for Biological Diversity (SP/CONAGESE, 1999). These documents are the basic reference documents for the implementation of the Convention.

During the World Summit on Sustainable Development held in Johannesburg in 2002, a breakthrough was made in commitments to environmental protection with the ratification of the treaty on the conservation of natural resources and biodiversity.

The Ramsar Convention on the protection of the environment, especially of wetlands, was adopted on February 2, 1971 by 170 signatory countries. It came into force on December 21, 1975. According to the texts of the Convention (Article 1.1), wetlands are: "areas of marsh, fen, peatland or water, whether natural or artificial, permanent or temporary, with water that is static or flowing, fresh, brackish or salt, including areas of marine water the depth of which at low tide does not exceed six meters" (SCR, 2013). In a bid to protect wetlands that encompass vast forest areas, Burkina Faso, after ratifying the said convention, designated fifteen (15) sites on the Ramsar list for a total area of 652,502 hectares (MEDD, 2013). In 2013, the country adopted Burkina Faso's National Wetlands Policy.

[3]http://www.un.org/fr/events/biodiversityday/convention.shtml Accessed on 7/9/2018.

Finally, one of the major agreements on environmental protection is COP 21 on climate, held in Paris at the end of which, the agreement provides to contain global warming "well below 2° C above pre-industrial levels and if possible to aim to continue efforts to limit the rise in temperatures to 1.5° C" .[4]

1.2. The problem

On a global scale, the number and extent of protected areas has increased significantly in recent years, raising the global protected land area to 12 per cent. Nevertheless, the variability and integrity of these protected areas are seriously threatened by direct and indirect interventions (SCDB, 2004). Today, they are often the only hope we have of preventing many threatened or endemic species from disappearing forever (Dudley, 2008). The management of protected areas is not always in line with the guidelines given their status by specialized bodies such as the IUCN.

Numerous shortcomings have been noted in terms of the development and implementation of effective management instruments for these protected areas. The status of most protected areas in West Africa is currently contested, in that they were classified during the colonial period. The management modalities of these classified areas have never been reviewed in line with the changing context and possible new management modes (IUCN/PAPACO, 2012). This failure to adapt the management of these protected enclaves to current realities affects their integrity. In developing countries, shortcomings in the management of protected areas are inherent to the problems of poverty and

[4] http://fr.wikipedia.org/wiki/Accord_de_Paris_sur_le_climat consulted on 20/7/2016.

demographic growth, insofar as forests and trees are the basis of the economy in most African countries (Chidumayo & *al.*, 2011). To a lesser extent, they contribute to poverty reduction among forest-dwelling populations (Somé, 2012), and are also used by local populations for food, energy, traditional pharmacopoeia, timber and services.

The Sahelian population is one of the fastest-growing in the world (Ozer, 2006). In Burkina Faso, over a period of thirty years (1975-2006), the population rose from 5,638,203 to 14,017,262, representing an increase of 148.and a multiplication by a factor of around 2.5 (Sangli & Tallet, 2012). The population explosion after the 1960s (growth rate of around per year) in most Sahelian countries has increased the pressure on natural resources (Bontioni & Reij, 2009).

In Africa, many protected areas are vulnerable to negative influences such as pollution, logging, illegal hunting and agricultural encroachment (Mengue-Medou, 2002). The unavailability of forest products and agricultural land is increasingly leading people to exploit protected areas. They are threatened with "nibbling" by farmers faced with a shortage of farmland (Gomgnimbou, 1998; Fournier & Millogo-Rosolodimby, 2007; Dimobé, 2017). The sharp increase in cultivated areas and the average decrease in fallow periods are disrupting the landscape (Fournier & al., 2001).

The cultivation systems implemented in these protected areas are therefore incriminated. Overexploitation of resources and inappropriate farming techniques are disrupting environmental dynamics (Hauchart, 2005). Shifting cultivation through frequent land clearing in protected areas is at the root of deforestation and the reduction in forest cover (Sounon Bouko & *al.,* 2007; Tankoano & *al.,* 2016; Kadéba & *al.,* 2019). It modifies vegetation structure (Tianhoun, 2012). The reduction in forest cover affects not only flora and fauna, but also agriculture itself (Solly & *al.,* 2020).

The most striking consequences of deforestation are the increase in runoff (Mahamane, 2006), and consequently water erosion of soils. Anthropogenic actions (logging, clearing, pastoral pressure, etc.) are generally considered to promote the sealing of surfaces and consequently the accentuation of runoff (Aurouet & *al.,* 2005). Soils under plant formations are more resistant to water erosion than soils under cultivation and bare soils (Yaméogo & *al.,* 2020). We are therefore witnessing a significant loss of arable land (Avakoudjo & *al.,* 2014). Farmers are forced to extend their areas to increase agricultural production by clearing new fields, generally in protected areas but also in marginal spaces. Sensitive areas, notably riverbanks (Daou & *al.,* 2019), are also exploited through expansive farming methods, especially during rainfall deficit years. These practices increase the risk of degradation and pollution of these areas (Kaboré, 2013).

Agricultural clearing is responsible for the decline in floristic diversity and density. This degradation is accelerated by the resurgence of cotton cultivation in protected areas (Dipama, 2009).

Woodland formations and protected areas are used as the main sources of grazing in Sahelian countries with large herds of livestock. Pastoral practices are diverse and varied. Cattle graze shrubs and trees, while herders climb trees such as *Pterocarpus erinaceus*, *Afzelia africana* Sm.

and *Detarium* and *Acacia* species to cut / prune foliage to provide extra feed for their livestock during the dry season (Sawadogo, 2007; Muoghalu & *al.,* 2014). Overgrazing is therefore a serious problem in the Sudano-Sahelian strip, where overcrowding causes severe damage to vegetation (Muoghalu & *al.,* 2014). Selective cutting of plant species, especially in protected areas, contributes to the erosion of biodiversity. Increased grazing pressure favors the progression of certain species (*Guiera senegalensis*, *Eleusine indica*) and the regression of forage species (*Acacia albida, Khaya senegalensis*, *Ziziphus mauritania*, *Andropogon gayanus*, *Andropogon ascinodis*) (Devineau & Fournier, 1998).

Livestock farming reduces or even prevents the natural regeneration of vegetation, although some authors classify it as a secondary cause of vegetation degradation (Ariori & Ozer, 2005). It has been reported that of woodland degradation is due to overgrazing, which is concentrated in semi-arid zones (Chidumayo & *al.*, 2011). Some shrubs are destroyed by trampling and grazing. Lack of pastoral space and excessive animal numbers are the main causes of illegal use of protected areas (Kiéma & *al.,* 2007), which are degrading insidiously. The obstruction of cattle tracks by fields leads to numerous conflicts between farmers and herders. Herds encounter enormous difficulties in accessing resources, notably water and grazing (Palé, 2015). Protected areas then become attractive for pastoralism, not only because of their pastures but also because of their absence of fields (Boutrais, 2008).

Vegetation monitoring (floristic composition, biomass, area) can be coupled with other variables of a natural (climate, erosive processes) or social (pastoral or forestry practices) nature (Cohen, 2008). The droughts of the 1970-1980 decade and of 1984 and 1985 in the Sahel had a particularly damaging impact on water resources and vegetation. While rainfall, temperature and radiation are necessary for vegetation growth at a given point on the planet, and terrestrial ecosystems indirectly influence climate through their atmospheric exchanges (Ago, 2016). In West Africa, climate exerts a crucial influence on vegetation and settlement (Ballouche & Rasse, 2007; Orsenna & Petit, 2011).

The impacts of climate change on this part of the African continent are characterized by an increase in climatic extremes, the spread of desertification and the loss of biological diversity (Dorsouma & *al.*, 2008). Drought is expressed not only in terms of rainfall deficit, but also in terms of interannual variability, which can exceed (Zoungrana, 1995). Notwithstanding the relative improvement in rainfall described and predicted by many authors (Ozer & *al.,* 2010; Ouoba, 2013), it has been accompanied by a significant reduction in the number of rainy days and a shortening of the rainy season (Ozer, 2005). Climatic deterioration, from the Sudanian to the Sahelian zone, leads to a reduction in density, in the number of plant species and in the families to which they belong, in tree and shrub strata (Pallo & Sawadogo, 2010).

High plant mortality rates in some protected areas are sometimes attributed to climate change (Adjonou & al., 2009), notably rising temperatures and falling rainfall. Water remains the essential element for plant germination and growth in the Sahelian zone. In line with Diouf & *al* (2000), vegetation, along with soil and water, is one of the main elements whose dynamics influence the evolution of the forest ecosystem. Without water, plantations and seedlings, whether natural or artificial, have no future in dry zones (Bellefontaine & *al.*, 2003). Thus, low rainfall levels and their poor distribution in time and space contribute considerably to the fragility of these ecosystems (Diello & *al.*, 2005). These variations in certain essential climate parameters lead to changes in the structure and even floristic diversity of plant formations. Regressive climate variability, manifested through extreme weather phenomena such as high winds, floods, sandstorms and high temperatures, not only limits plant growth but also influences the dynamics of natural resources (Ouoba, 2007; Ngo & *al.,* 2018).

The degradation of protected forest areas and other woodland formations, the main refuge for wildlife, is a source of threat for animal species, some of which are in danger of extinction. Of the world's 2,970 endangered animal species, Africa has the highest number of species listed as "critically endangered" and "endangered" (Jacquemot, 2018).

Protected areas in Burkina Faso are subject to strong pressures from a variety of sources. However, cultivation is by far the main factor in the anthropization of forests (MECV, 2007). In fact, the area under cultivation in Burkina Faso has increased significantly. Agricultural activities as a whole occupied 29.of land in 1961, compared with 41.in 2007 (Cambrezy & Sangli, 2011). In addition, the average annual rate of expansion of the agricultural domain between 1990 and 2014 was 0.61% for an area of 163,326 ha (MEEVCC, 2019).

The colonization of protected areas by agriculture has led to a reduction in forest area. Protected areas are located in village terroirs, where local logics do not take into account the fact that these areas are no longer part of their "traditional sovereignty domains", but belong to the State or to external stakeholders (Valéa, 2010). Over the period 1992-2002, all forest formations decreased by 107,626 ha, representing an average degradation rate of 0.83% (MEEVCC, 2019). In 2002, the country's natural forest formations (open forest, gallery forest, shrub savannah, tree savannah, steppe) covered 13,305,238 hectares, or 48.of the national territory (MECV, 2007a; MECV, 2007b).

The State's classified forest estate covered an estimated total area of 3.9 million hectares, or 14% of the national territory. According to MEDD (2011), in 2010classified areas covered just 12% of the territory, well below the required internationally, and the annual rate of loss of forest formations between 1990 and 2010 is estimated at 0.87% for a total annual area of 59,900 ha. The total loss of forest cover in 20 years is estimated at 17., representing a total area of 1,198,000 ha. According to the same source, of the 69 protected areas, two (2) are national parks (390,500 ha), fourteen (14) are wildlife reserves (2,545,500 ha) and fifty-three (53) are classified forests (880,000 ha). The living biomass of these forest formations contains the equivalent of 292 million tonnes of carbon.

Wood fuel is also a major factor in the analysis of deforestation and biodiversity erosion in Burkina Faso. Indeed, low overall household incomes force households to exploit wood as their main source of energy. In 2008, wood accounted for 74.of the overall energy balance, and 94.of total household energy consumption (MECV, 2010).

Continuous deforestation reduces the floristic richness of forest species (Ozer, 2005). There is inevitably a significant loss of biodiversity when forests disappear (De Foresta, 2008), as they provide a habitat for many species of fauna (wildlife and microfauna) and flora.

In recent decades, deforestation in the Sudano region of Burkina Faso has been accentuated by a dynamic agricultural front driven by significant population growth and migratory flows to the west and south-west. These migratory waves were initially organized by the State, which had developed agricultural areas to settle drought-stricken populations within the framework of the Volta Valleys Development Authority (AAVV). But some populations are migrating independently to this area, which is ideal for farming, pastoral and gold mining activities. In savannah or desert regions, along watercourses, farmers, herders and fishermen often settle (Dipama, 1997).

The classified area of the South-West region includes wildlife reserves and classified forests covering some 153,700 ha. These protected areas, the main reservoirs of biodiversity in the area, are now threatened by climate variability combined with various anthropogenic pressures. Studies carried out in the classified forests of this region indicate a high level of anthropization resulting in a decline in woodland formations (Tankoano, 2012; Somé, 2013; Dimobé, 2017, Korgo, 2018).

Like almost all classified forests in the South-West, Bontioli's wildlife reserves are subject to heavy pressure from a variety of sources, characterized by over-exploitation of resources. Indeed, the lack of a clear management plan for these protected areas paves the way for the depredation of their resources.

The impact of climate change and population growth in the central and northern parts of the country, combined with the Ivorian crisis of 2002, have led to significant waves of migration into the area. In addition, artisanal gold mining has intensified in these reserves over the last two decades, attracting more and more people. The ecological impact of this activity is immeasurable.

The wildlife habitat is therefore gradually disappearing, even though the main reason for classifying these forests was that they were part of a tourist and hunting circuit passing through Diébougou, Gaoua, the Bouna reserve (Côte d'Ivoire) and Banfora (IUCN/PACO, 2009). They should provide effective protection for game. Like the other classified forests, the Bontioli wildlife reserves are of ecosystemic and socio-economic interest. They represent one of the most important classified forest entities in the region. Not only do they contribute to the sequestration of atmospheric carbon, the water cycle and the maintenance of biological diversity, they also provide a livelihood for local populations. The Bontioli Total Wildlife Reserve is one of the country's four (04) total reserves specifically dedicated to the integral protection of wildlife. In addition to protecting and preserving usable natural resources or providing environmental services, protected areas are expected to generate income (hunting, eco-tourism, recreation, and jobs for managers, guides, operators, etc.) (De Boissieu *& al.*, 2007).

Notwithstanding the usefulness of these protected areas, very few studies have been carried out on the dynamics and management of their plant formations. It should be noted, however, that there is no comparative study based on the status of each reserve (total and partial). To fill these gaps, and in view of the various pressures they are continually subject to, it seems important to take a critical and comparative look at the evolution of plant formations in terms of spatial coverage and floristic composition, as well as their management methods.

To do this, the main question is how are the plant formations in the Bontioli wildlife reserves evolving? specifically:

1. What are the ecological characteristics and spatio-temporal dynamics of plant formations in the reserves between 1986 and 2018?
2. How can we explain the impact of natural factors on the evolution of plant formations?
3. How have socio-economic factors influenced the evolution of plant formations in the reserves?
4. How are Bontioli's wildlife reserves managed?

Based on these research questions, hypotheses and objectives were formulated to give a precise direction to the study.

1.3. Assumptions and objectives

Vegetation is an environmental component whose dynamics depend on its interaction with other environmental components. A modification of the physical environment in a climax situation will lead to a new evolution in a new climatic position. Thus, a structural change in climatic conditions, edaphic conditions or fauna can have an impact on both the vertical and horizontal structure of the vegetation, as well as on its floristic composition. The same applies to a change in human populations, their systems of representation, forms of spatial organization and public policies.

The context of the Bontioli reserves has changed. This evolution affects both the physical and the human environment. This is why, from a theoretical point of view, it makes perfect sense to consider the evolution of vegetation too.

1.3.1. Research hypotheses

The main assumption is that plant formations in reserves evolve in a jagged pattern under the influence of natural and socio-economic factors.

From this main assumption derive the following specific assumptions:

- The plant formations on the reserves have a low floristic richness and diversity and have undergone significant changes between 1986 and 2018;
- The main climate parameters have a negative impact on the evolution of plant formations in the reserves;
- Inappropriate farming practices have had a negative impact on the evolution of plant formations in the reserves;
- Local communities are not sufficiently involved in reserve management.

From these hypotheses, the objectives of this research were developed.

1.3.2. Research objectives

The main objective is to analyze the evolution and management of plant formations in the Bontioli wildlife reserves.

Specifically:

- To describe the ecological characteristics and spatio-temporal dynamics of plant formations in the reserves between 1986 and 2018 ;

* Explain the influence of natural factors on the evolution of plant formations;
* Examine the impact of the main socio-economic determinants on the evolution of plant formations;
* Analyze reserve management methods used by local residents.

1.4. The state of the art

For a better understanding of the evolution of living beings (animals and plants), it would seem advisable to set out a theoretical orientation for this study.

1.4.1. Biogeography as a field of investigation

The term biogeography, originally plant geography, according to the Petit Larousse de la langue française, refers to the science of studying the distribution of living beings on the earth's surface, as well as changes in this distribution. Biogeography is a branch at the crossroads of the so-called natural sciences, physical geography, pedology, ecology, bioclimatology and evolutionary biology, which studies life on the surface of the globe through descriptive and explanatory analyses of the distribution of living beings, and more specifically of communities of living beings[5] . Its aim is to describe, explain and model the spatial distribution of living beings (Cohen, 2008). For Velle & Zandée (2002), it is the scientific discipline that attempts to explain the distribution of groups of organisms, or taxa, on the earth's surface.

The theme of this research falls within the field of biogeography, in terms of the object pursued. The clarification of a theoretical frame of reference on which to base the study seems vital, and has a dual interest.

---

[5]https://fr.wikipedia.org/wiki/Biogégraphie consulted on 3/4/2019

Not only does it provide a better framework for analysis, but it could also help users to better understand the results of the study.

1.4.2. Biogeography: evolution and links with other sciences

For over two centuries, biogeography has been struggling to find its feet. It has taken several names: "botanical geography", "geography of plants and animals", etc., but the term only appeared at the end of the [19th] century. Marie-Claire ROBIC points out that it was as early as 1890 that the word was used by French geographers when they discussed the work of their German colleagues (Drouin, 1995). The term "biogeography" was first used in 1888 by Friedrich Ratzel. But the ambition of the biogeographer is, at the very least, to describe the distribution of taxa on the surface of the globe and, at best, to write their history, without excluding the hope of predicting their future (Deleporte, 2002).

The history of biogeography as a scientific discipline in the making has been shaped by a number of theories.

1.4.2.1. Major biogeographical theories

Numerous theories have furthered biogeographic research. For the purposes of this study, we will focus on those that have had the greatest impact on the evolution of this discipline.

1.4.2.1.1. The theory of Carl von Linné (1707-1778)

The first coherent biogeographical theory of this kind was formulated in the eighteenth[th] century by Linnaeus (Cecca & Zaragueta, 2015). Based on a biblical interpretation of creation, Linnaeus accepted a single origin for living beings, the "paradise" from which dispersal took place according to environmental conditions (temperature, humidity, topography, nature of the soil, etc.). This discontinuity in regional natural geographic phenomena would be at the origin of the colonization of the globe by species of different natures. This theory was not empirically testable, especially as regards the common origin of species. Notwithstanding, the author would have had the merit of being renowned as the creator of the binominal system almost universally used today to name different groups of living beings and fossils, arranged in nomenclatural categories (species, genera, families, order, etc.) (Cecca & Zaragueta, 2015). The failure of Linnaeus' theory to be empirically tested has inspired other authors to use other elements to develop more coherent, field-testable theories.

1.4.2.1.2. The theory of Georges Louis Leclerc, comte de Buffon (1707-1788)

Buffon's theory is based on the Linnaean postulate, which he tried to explain by adding new elements after studying the distribution of mammals in tropical Africa and South America. He incorporated migration-related mutations into his theory. For him, during migration, living species are confronted with conditions that are often hostile to them, and this can lead to changes within these species. Buffon used the term "adaptation" to explain differences in species' tolerance of abiotic conditions (Morin, 2006).

He established one of the first principles of biogeography, known as Buffon's law: "distinct assemblages of species occupy isolated but environmentally similar regions". Buffon's theory was not without its critics.

1.4.2.1.3. The theory of Charles Darwin (1809-1882)

Darwin built on the theories of his predecessors, in this case Buffon, by integrating a new notion: evolution. His theory of evolution and species adaptation was a spectacular success. The foundation of his thinking was laid by observations made on birds from the Galapagos Islands, during which he observed the colonization of oceanic islands by continental species, and the mutation of the latter over time. It should be noted that, at the time, almost all naturalists supported the thesis of a common origin of living beings, inspired by biblical writings.

Authors such as Ernst Haeckel (1834-1919), inventor of the term "ecology", was a fervent defender of Darwin's theory.

The 20th century marked a decisive turning point in the evolution of biogeography. Wegener's theory of plate tectonics and continental drift (1912) provided additional arguments to explain the distribution of species across the globe.

Ultimately, the spatial distribution of living organisms is the result of the influence of both internal and external factors (Somé, 2010).

1.4.2.2. Biogeography: methods, approaches, links with other sciences

Its authorship is disputed by many disciplines. For some, it belongs to the biological science encompassed by ecology. Others, on the other hand, believe it derives from a branch of geography. Finally, some consider biogeography to be an integrative science, combining parts of different disciplines that have already been established: geography and ecology, but also botanical and zoological taxonomy, paleontology, geology, population genetics, etc. (Drouin, 1995). It transcends traditional disciplines (Hugot, 2002).

Drawing on recent developments in the techniques of cartography, remote sensing and Geographic Information Systems (GIS) in scientific research, this discipline is at the heart of global concerns for environmental management and protection. As it evolves, questions of a political nature also come into play. Public policy is undoubtedly the driving force behind the dynamics of plants, animals and, in short, biodiversity.

According to Drouin (1995), there are two main lines of research. Modern biogeography comprises two fields: historical biogeography, which seeks to reconstruct the origin, dispersal and extinction of taxa and biomes, and ecological biogeography, which seeks to understand present distributions and geographical variations in diversity in terms of the interaction between organisms and their biotic and abiotic environments (Morin, 2006).

The biogeographical approach is not uniform. It can be systemic, ecological or physiognomic, depending on the objectives pursued. Increasingly, the systemic approach is being favored, as it enables us to take into account several dynamically interacting environmental elements in the study of a given phenomenon. It's a multi-disciplinary approach that aims to consider all related dimensions in the analysis of a phenomenon. For example, the study of plant dynamics would not be complete or convincing without taking into account the social (anthropic), pedological, climatic and even morphological aspects. The aim is therefore to deepen the study in order to achieve scientifically satisfactory results. The systems approach is by no means a panacea for research, but it does have its credentials in the humanities, social sciences, biology and so on.

1.5. Clarifying concepts

Forest: the origin of the word "forest" comes from the Latin *forestis,* itself derived from *foris*, meaning "outside". Forests therefore refer to vast areas beyond human influence, habitat or cultivation.

Montagne & *al.* (2005) point out that 18th-century metropolitan foresters defined the forest as a "separate space devoted to timber production and soil protection".

A forest is not just a more or less extensive and diversified group of trees, but a complex ensemble of living beings, animals and plants, closely associated, but some of which have a more direct interest than others for the people of the moment (Guillard, 2014).

A distinction is made between dense or closed forests and open or open forests, depending on whether the tree crowns are contiguous in the former case, or discontinuous in the latter. In Burkina Faso, closed forests are formations where the different levels and undergrowth cover a very large portion (over) and have no herbaceous stratum. (...) gallery forests are considered closed forests. Open forests are formations where trees are present in a discontinuous way with a cover of at least and maximum. The various types of savannah, with the exception of grassy savannah and open forests, constitute open forests (Ouédraogo, 2001).

Whereas the FAO[6] defines forest as corresponding to tree cover of more than over at least half a hectare. A tree is defined as a perennial plant with a single stem (or several if it has been cut back) reaching at least five meters at maturity.

In legal and regulatory terms, there has been an evolution in the definition of the term "forest". During the colonial period, the protection of the forest had purely economic aims, rather than the conservation of biological diversity. This is how the concept was defined in article 2 of the decree of July 4, 1935: "Forests are defined as lands whose exclusive or main products are wood for cabinetmaking, industry or services, firewood and charcoal, or accessory products such as bark and tannin fruits, textile and dye bark, kapok, rubber, glue, resins, gums, bamboo, wild palms and all other plants that do not constitute an agricultural product" (Kiéma, 2007). This definition corresponds to the legal one at the time.

According to article 10 of the Burkina Faso Forestry Code (2011), forests are lands with plant formations based on trees or shrubs and grasses, excluding those resulting from agricultural activities.

Obviously, for the sake of consistency, the definition used is that of the Burkina Faso Forest Code, which ignores the parameters of density and extent of vegetation cover.

Ecosystem: the term ecosystem was coined by Arthur George Transley in 1935 to designate the basic unit of nature.

The ecosystem is the basic structural unit of the biosphere. It is made up of biotopes forming a biocenosis. It is homogeneous in terms of

[6]http://www.fao.org/docrep/003/X4130f4130f04.htm#Top0fPage consulted on 8/10/2014.

topography, microclimate, botany, zoology, hydrology and geochemistry (George & Verger, 2009).

For Da & Yonkeu (2007), an ecosystem is a relatively homogeneous and stable environment in which all living beings maintain food and territorial relationships with each other and with the environment.

Generally speaking, an ecosystem refers to all the interrelationships that exist between the components of a single system in a given environment (physical, political, legal, cultural or economic). Thus, we speak of economic, political, etc. ecosystems. However, for the purposes of this study, the term ecosystem is used in the sense given to it by the above-mentioned authors. It is thus defined as an ecologically dynamic space made up of a set of biotic and abiotic environmental elements in permanent interaction so as to create a balance for optimal functioning of each of the system's components and of the system as a whole.

Landscape: there is no a priori universal description of a geographical space or object, since the description comes from the observer according to his or her objective, with a great perception of geographical diversity (Bigot, 2004). For him, the landscape is a story, meaning that at a given moment, the possible evolutions depend on the instantaneous state, but also on the information accumulated in the past, i.e., the initial conditions.

Landscape is the result of a continuous confrontation between society and its environment (Bamba, 2010). It is also the result of a certain portion of space, the dynamic and therefore unstable combination of physical, biological and anthropological elements which, through dialectical reaction between them, create a unique and indissociable whole in permanent evolution (Bertrand, 1968).

Landscape is a complex concept; it refers both to the natural environment, free from any human influence, and to a socially constructed space. In the latter case, man is the main actor in this geographical space, which he creates according to his activities, his technology and his perception of the natural environment. In any case, the notion of landscape cannot be separated from that of geographical space with tangible physical limits. Sanguin & Pĕna (1986) argue that landscape has a spatial connotation that ecosystem, a fundamentally biological concept, does not possess, even if in some cases the two concepts can be assimilated.

Biological diversity: the expression was coined by American biologist Thomas Lovejoy in 1980. It was taken up again in 1985 for the National Forum on Biological Diversity by biologist Walter G. Rosen, who coined the neologism biodiversity. A catchword for the heterogeneity of the living world and the interactions that govern it .[7]

According to UNCBD (1992), it refers to the variability among living organisms from all sources, including, inter alia, terrestrial, marine and other aquatic ecosystems and the ecological complexes of which they are part; it includes diversity within species, between species and of ecosystems.

According to Gramond (2010), biodiversity results from the potential of :

[7]http://environnement.doctissimo.fr/faune-et-flore/

- Intraspecific convergence (spatial and temporal coexistence-competition mechanisms),
- Spatial and/or temporal interspecific differences (competitive exclusion).

Biodiversity therefore covers three levels of variability in the living world: within living species (or genetic diversity), between species (interspecific diversity) and between ecosystems (ecological diversity). Alongside an inventory of the various components of biodiversity, this approach also emphasizes the notion of "interactivity" between these three different levels of organization . [8]

According to Lepart (1997), the term "biodiversity" refers to the variety of living organisms, whatever their environment of origin, and takes into account intraspecific, interspecific and functional diversities.

Biodiversity is sometimes referred to as the "resource of resources", as it is the primary source of many vital resources such as air, drinking water and everything we eat, as well as a large part of our energy resources (fossil fuels and biomass), medicines, textile and paper fibres, and so on.[9]

Today, the term biodiversity often refers to the species richness of an ecosystem. Its measurement is then limited to a taxocene, i.e. a subset of the species of a taxonomically related community: butterflies, mammals, trees (Marcon, 2018). In this work, biodiversity concerns floristic richness in the plant formations of reserves.

---

[8] https://www.thegef.org consulted on 14/4/2018

[9] http://fr.wikipedia.org/wiki/Ressource_naturelle consulted on 5/11/2017

Plant formations: a plant formation is a community of plant species that, along with other elements, determines a characteristic landscape, which is often given a particular name by the people who frequent it or the scientists who study it .[10]

In botany or biogeography, a plant formation designates a classification of plant species characterized by a particular aspect which, in turn, determines a characteristic landscape. It specifies the nature of a group of plants with a similar habitus, forming a homogeneous plant unit .[11]

Vegetation types depend on the degree of cover. Based on the classification of African vegetation types adopted at the Yangambi conference (Aubreville, 1957) and according to Tankoano (2012), we distinguish :

- forest gallery: 90% to cover ;
- open forest: to cover, with some shrubs ;
- wooded savannah: 30-60% cover with some shrubs;
- wooded savannah: 30-60% cover with sufficient shrubs
- shrub savannah: 15-30% cover, shrubs are dominant, but trees are rare.

The distinction between tree savannah and wooded savannah is that in tree savannah, woody plants are "scattered", while in wooded savannah they form "a generally clear canopy" (Descoings, 1978).

Plant formation is used in this research as a group of plant communities in a given geographical area, made up of floristic and herbaceous

[10]https://fr.wikipedia.org/wiki/Liste_des_formations_vegetales consulted on 15/5/2019
[11]https://www.aquaportail.com/definition-9830-formation-vegetale.html consulted on 15/5/2019

species, subject to identical or similar edaphic, topographical and climatic conditions, and presenting a homogeneous overall physiognomy.

Protected areas: areas specifically intended for the protection and maintenance of biological diversity, and of natural and associated cultural resources, the management of which has been envisaged by legal or other effective means (FAO, 2010).

A protected area is a clearly defined geographical space, recognized, dedicated and managed by any effective legal or other means to ensure the long-term conservation of nature and its associated ecosystem services and cultural values (IUCN, 1994). There are six (6) IUCN protected area categories (see Table 3).

Table3 : IUCN protected area categories

Category	Type
Category I	Strict nature reserve / Wilderness area
Category II	National parks
Category III	Natural monuments / Natural highlights
Category IV	Protected areas for habitat and species
Category V	Protected landscapes and seascapes
Category VI	Protected areas of managed natural resources

Source: IUCN, 1994

For Ly (2001), protected areas are spaces that benefit from special measures to protect and manage wildlife.

Depending on the objective pursued, there are several types of protected area: strict nature reserves, wildlife reserves, wildlife sanctuaries, biosphere reserves, and so on. Forest areas within a designated protected area can have the status of forest protected areas if the main objective is biodiversity conservation (Barbara, 2012).

Strictly protected areas set aside to protect biodiversity and possibly also geological/geomorphological features, where visitation, use and human impacts are strictly controlled and limited to ensure the protection of conservation values. These protected areas can serve as vital reference areas for scientific research and monitoring (Dudley, 2008).

In general, these are areas set aside to allow natural factors to operate freely without any external intervention, with the exception of the safeguards necessary for their very existence. All harvesting and other forms of exploitation (forestry, agriculture, mining, etc.) likely to harm or disturb the flora and fauna are prohibited (Ly, 2001).

"Total wildlife reserves are established for the protection of all wildlife species and their habitats; hunting activities are prohibited there" (article 94 paragraph 1) (Burkina Faso Forest Code, 2011). According to article 95 of the same code, the creation of total wildlife reserves is a matter for the law. While human activities are totally banned, fishing is prohibited and human presence is forbidden, with the exception of reserve staff and scientific missions with special authorization (IUCN, 2010). There are four (4) total reserves in Burkina Faso: the Arly total wildlife reserve in the east, the Madjoari total wildlife reserve in the east, the Bontioli total wildlife reserve in the southwest and the Singou total wildlife reserve in the east. They are classified in category I of the IUCN list.

Partial reserves: according to article 94 (paragraph 2) of the Burkina Faso Forest Code (2011), partial wildlife reserves are established for the special protection of certain species and their habitats; hunting activities are authorized. Article 95 specifies that partial reserves are established by a decree issued by the Council of Ministers. They are classified in category IV of the IUCN list.

Conservation: according to the Larousse dictionary, conservation means the action of preserving, keeping intact or in the same condition.

Synonymous, in principle, with preservation, protection, safeguard. Conservation applied to nature involves management designed to preserve one or more species of flora and fauna, or even a landscape, considered to be under threat (Giraut & *al.,* 2004).

Deforestation: defined as the elimination of at least of forest cover over a given area (Beaux, 2011). It is a dynamic in which forest is replaced by another type of land cover (Bamba, 2010).

Deforestation is the regression of forested areas. It is the result of deforestation and subsequent clearing, linked to the expansion of agricultural land, excessive exploitation of certain forest species and urbanization . [12]

For FAO (2010), the concept of deforestation refers to the conversion of forest to another land use or the long-term reduction of tree cover below a threshold of ten percent. Deforestation also refers to areas where, for example, the impact of disturbance, overexploitation or changing environmental conditions affect the forest so much that it cannot maintain a tree cover above the ten percent threshold.

Ultimately, deforestation is the progressive loss of the constituent elements of a forest or woodland formation (in savannah landscapes),

[12] www.conservation-nature.fr/article2.php?id=136 consulted on 27/11/2015

so that this vegetation unit can no longer perform its original functions, i.e. maintaining and producing biological diversity, storing carbon, maintaining the water cycle, etc. It occurs under the combined effects of destructive forces of various origins (anthropogenic and natural). This is due to the combined effects of destructive forces of various origins (anthropogenic and natural).

Forestry policy: the term "policy" is not precisely defined, but is used in different ways on different occasions. Often, it can mean "a plan of work adopted and pursued". A policy may or may not be explicitly stated. It can also be planned or emanate from obvious behavior. This is why it is often considered either as a rational system based on deliberate objectives and plans, or as the consequence of political activity resulting from a series of decisions (...). National forest policy is seen as a negotiated agreement between government and stakeholders (i.e. all those who depend on or benefit from forests, or who authorize, control or regulate access to their resources) on the orientations and principles of the actions they adopt, in line with national socio-economic and environmental policies, to guide and determine decisions concerning the sustainable use and conservation of forest and tree resources for the benefit of society.[13]

Forest policy is defined as the set of orientations and principles for action adopted by public authorities in harmony with national socio-economic and environmental policies in a given country, and intended to guide future decisions concerning the management, use and conservation of forests and trees for the benefit of society (FAO, 2010).

[13]File:///D:/i1679f03.pdf consulted on 9/10/2014

For Buttoud & Samyn (1999), forestry policy is defined as the set of standards and measures taken by a country's public authorities to regulate the use of forest resources and to add value to all products extracted from forests.

Burkina Faso's National Forestry Policy is defined as the overall reflection of the objectives it pursues, the options it takes into account, and the strategy and operational approaches that underpin the activities programmed within its framework. By virtue of its specific nature, it is supplemented by legislative texts that guide and facilitate its application, as well as administrative and technical management mechanisms and instruments. Its objectives are to:

- translate the place and role of the forestry-wildlife and fisheries sub-sectors, and specify the government's options through the related development priorities in line with the missions of the supervisory ministry;

- rationalize the management of resources in the three sub-sectors of forestry, wildlife and fisheries through the coherence and synergy of interventions in the three sub-sectors with clear improvements;

- to provide a conceptual basis for the drafting of legislation relating to the management of these three sub-sectors;

- provide a negotiating tool and a frame of reference for consultation with development partners to coordinate and harmonize interventions in the three sub-sectors (MEE, 1998).

The agrarian system: this refers to all the practices and techniques used by a society to manage rural areas for agricultural, pastoral and forestry production. This concept "encompasses the mode of exploitation and reproduction of one or more ecosystems, the social relations of production and exchange that have contributed to its establishment and development, and the overall economic and social conditions, in particular the system of relative prices that determine the terms of its more or less extensive integration into the world market" (Cochet, 2007).

Agrarian systems change with technological, social and economic developments. It can be assumed that, like all other agrarian civilizations in the world, those that have existed in this part of West Africa have evolved from the hunting and gathering age to the current stage (Ouédraogo, 2015) characterized by the gradual modernization of agricultural and pastoral production. The functioning of these agrarian systems is highly complex. Agrarian systems are always on the move. To grasp and understand their structure and functioning, we need to be able to define a "state" of the system at a given moment in its history, i.e., to imagine it stable, long enough to glimpse the interactions and fundamental mechanisms that characterize it. This is an essential step in perceiving and interpreting movement, in order to detect the conditions of its sustainability, or on the contrary, the causes of its impending crisis (Cochet, 2011).

The agrarian system is therefore determined by a set of human components, in particular the social group occupying a given geographical area, through its knowledge and know-how, and the natural components, in this case the land as an element of production, water, vegetation, and so on.

The agropastoral production system: this concept encompasses various notions, some relating to agronomy and others to economics, sociology and rural geography.

An agricultural production system is a way of combining land, labor and means of work for plant animal production, common to a group of farms. A production system is characterized here by the nature of the production, labor power (skills) and means of labor used, and by their proportions (Reboul, 1976).

A farm's production system is defined by the combination (nature and proportions) of its productive activities and means of production (land, capital, labor). The study of production systems includes the study of production sub-systems (breeding, cultivation and processing) which are characterized by the nature of the products, the technical itineraries followed and the yields of these productions. The evolution of production systems can be characterized by several main parameters: diversification / specialization (greater or lesser diversity of productions), intensification / extensification (in labor, capital or input per unit area) (Tristan & *al.*, 2009).

For Broussier (1987), the production system is the result of several combinations and refers to the farm as a whole: a particular arrangement of production factors, technical choices, a combination of productions.

This concept is defined by Sebilotte (1974) as a set of technical itineraries, i.e. ordered and dated successions of cultivation techniques and practices applied to cultivated plant species with a view to obtaining products that are sold or transferred.

In this study, agropastoral production systems refer to all the cultivation and pastoral techniques and practices used by a rural community, established in a given geographical area, to satisfy its food, economic and cultural needs.

Ecotourism: ecologically oriented tourism (Worou & Sinsin, 2009).

According to The International Ecotourism Society (Ties) (Wallace, 1998), it is defined as travel to natural areas that conserve the environment and maintain the well-being of local populations.

In this work, ecotourism refers to all forest management activities and tourist infrastructures (rest areas, restaurants, accommodation, etc.) geared towards ecological tourism. Ecotourism is booming in savannah ecosystems, given their abundant wildlife. These ecosystems, which are still home to a number of endangered animal species that are fully protected by law (African elephants, lions, hippos, crocodiles, panthers, etc.), are attractive environments for tourists.

1.6. Demographic theories and the evolution of natural resources

Population growth has haunted the minds of rulers from the second millennium BC to the present day. Plato, who feared overpopulation of the city, argued for demographic fixity of settlement and presented a whole set of measures to adjust the population. For him, births should be reduced authoritatively, if necessary (Dumont, 2014). Aristotle takes up Plato's idea, but his concern for law and order, an essential principle, means that settlement must not undermine it. He set a marriageable age and required abortion if necessary. Efforts to regulate the birth rate have continued and been stepped up in various parts of the world.

1.6.1. The Malthusian theory of population growth

Thomas Robert Malthus, an English economist, published his landmark Essay on the Principle of Population in 1798, in which he made two major observations:

- food is necessary for human existence;
- reciprocal passion between the sexes is a necessity and will remain much as it is now.

In other words, he writes that the multiplying power of population is infinitely greater than the power of the earth to produce man's subsistence. He asserts that population increases geometrically, while subsistence increases arithmetically. Hence the need to balance these two powers. Malthus identified several mechanisms for regulating population, which he called "brakes". Grouped into two categories, he distinguished between active brakes: mortality resulting from population excess, wars, excesses and many kinds of avoidable evils, and preventive brakes composed of "moral constraint" and "vice". For him, moral restraint is abstinence from marriage and chastity. As for vice, it's libertinism, promiscuous sexual relations, "passions contrary to nature", i.e. homosexuality, "desecration of the marriage bed", i.e. "adultery", and all the devices employed to conceal the consequences of criminal or irregular liaisons, i.e. birth control and induced abortion.

The relationship between man and nature is thus analyzed in terms of human density, the extent of which could dangerously compromise the equilibrium of the ecosystem. In the spirit of this theory, human society is seen as a "weapon of mass destruction" of natural resources. For Demont & Jouve (1999), the Malthusian school assumes that competition for increasingly scarce resources (land, water, mineral resources, etc.) leads to poverty, degradation of the biophysical environment, conflict and a reduction in the rate of increase in population income. So, why set out Malthus' theory in this research work? How can we bring it back and analyze it at the local level?

The cross-disciplinary nature of this study of plant dynamics means that a number of parameters need to be taken into account. Indeed, population has always been perceived as the main user of natural resources, hence the need to set out the Malthusian theory that established the relationship between demographic evolution and the depletion of natural resources. With this logic in mind, the author ignored the technological evolution linked to the intensification of agricultural production, even though his country, England, was already in the midst of an agricultural revolution driven by scientific and technical progress. However, this theory is still relevant today, given the technological "backwardness" observed in some parts of the world, and the inefficiency of traditional resource-use techniques in the face of accelerating population growth. It is in this context that it seems appropriate to identify similarities between this general theory and the realities of the Bontioli wildlife reserves and their surrounding villages.

1.6.2. Neo-Malthusian theories

Malthus's theory has had immeasurable success. It was taken up and applied in various fields. Population adjustment or "anti-natalist" policies have been practiced in China and other countries around the world.

1.6.2.1. Climate Malthusianism

It's a crypto-Malthusian pro-climate trend. It originated on November 18, 2009, from the United Nations Population Fund (UNFPA), according to which the world's population should not exceed "an ecologically sustainable figure" (Dumont, 2014). Advocates of this movement believe that it is imperative to limit global population growth in order to "save" the climate. For them, it is the pressure on resources that is responsible for the degradation of climatic parameters.

1.6.2.2. The Malthusianism of natural resources

Since man is the main user of natural resources, this school of thought, derived from that of Malthus, argues that increasing population density leads to increased clearing of land for agriculture and domestic and industrial wood consumption, resulting in a decline in available natural resources, or even their irreversible disappearance.

1.6.2.3. The demographic dividend

The demographic dividend, a concept that emerged at the start of the 21st century, states that economic productivity rises when the ratio of the working population to the number of dependents increases. This policy of the international community aims to encourage developing countries, particularly those in sub-Saharan Africa, to begin the demographic transition. This is why it is accompanied by various actions aimed at young people, notably education and sexual and reproductive health.

In parallel with the proliferation of neo-Malthusian theories targeting mankind and its impact on natural resources, a new theory emerged in the second half of the ^{20th} century: that of Esther Boserup. This theory further enriched the scientific debate, while putting into perspective mankind's responsibility for the eventual disappearance of natural resources, as supported by its predecessor, which had not yet been a reality for over two centuries.

1.6.3. Esther Boserup's theory

In 1965, Esther Boserup was the forerunner of a current of thought opposed to that of Malthus. The theoretical foundation of her thinking is set out in her essay "Agrarian Evolution and Demographic Pressure", published in English in 1965 and translated into French in 1970. It is based on the premise that population growth leads to lower yields, as the fallow and cultivation periods shorten and lengthen respectively. This stimulates farmers to adopt techniques that allow more intensive occupation of the land. This evolution requires a higher labor input for field work and land management, but results in higher production per unit of labor (Demont & Jouve, 1999). For these authors, Boserup counters Malthusian pessimism by taking into account farmers' agronomic practices. She identifies several stages in the evolution of agrarian systems in relation to population growth. These stages range from slash-and-burn cultivation to irrigated cropping systems that can include several crop cycles in the same year (Jouve, 2004).

The author highlights mankind's intrinsic capacity to innovate in order to solve the natural resource deficit resulting from population growth, by inventing new agricultural techniques and tools to intensify production and, in turn, increase yields. All of which could help feed the world's ever-growing population.

Technology can certainly improve agricultural yields, but it should not be seen as a panacea for the management of natural resources. It often has irreversible adverse effects on the environment. Intensive irrigation aims to increase productivity, but can lead to salinization of the soil, leaving the land uncultivated and resulting in lower agricultural yields and population migration. The example of Mesopotamia, an ancient civilization of irrigated agriculture, is a case in point: of the soil in lower Mesopotamia is more or less affected by salts (Job, 1992). Certain products used in agriculture to improve yields could have harmful effects on human, animal and even plant health. Glyphosate, for example, is a carcinogenic insecticide. In developing countries with the lowest rates of school enrolment, technological advances in agriculture could have a downside in terms of the misuse of pesticides, herbicides and other agricultural inputs. This can lead to environmental disasters.

For Ouharon (2006), both authors (Robert Malthus and Esther Boserup) fail to take into account long-term threshold effects, irreversibilities, option values, substitution techniques and uncertainty regarding the future evolution of technical progress.

In this chapter, the study problem was presented, and the hypotheses and objectives formulated. This was followed by a review of the current state of knowledge on the subject, summarizing the literature on forest governance, the evolution of plant resources, agrarian systems and the effects of climatic variability on vegetation. Biogeographic and demographic theories related to the distribution and evolution of living beings and natural resources were also presented. For a better understanding of the study, concepts relating to the theme were clarified.

CHAPTER 2: THE METHODOLOGICAL APPROACH

The methodological approach adopted for this study consisted firstly in choosing the approach for conducting the work. In view of the objectives set, the best way to take into account a large number of data was identified. To this end, the study variables and indicators were described. Next, the spatial and demographic samples were determined. Finally, the techniques, tools and materials for data collection and processing were explained.

2.1. Which methodological approach for the study?

For a holistic view of reserve degradation, the systems approach has been adopted in order to conduct this research properly.

Originating some sixty years ago in the United States at the Massachusetts Institute of Technology, and applied in the development of the Tennessee River basin, the so-called systems approach is part of a philosophical current (structuralism) whose aim is to gather as much knowledge as possible about a given element with a view to greater efficiency of action (Da, 2005).

The system is defined as a set of dynamically interacting elements, so that if one element fails, the whole system is affected. This vision is appropriate for this research insofar as the landscape of protected areas is a set integrating elements of the physical environment (land, trees, water, wildlife, aquatic fauna, microfauna etc.), and human (populations and activities), forming a coherent whole. It's an approach that emphasizes the importance of interactions between the diverse phenomena that make up the specificity of an area (Dipama, 1997). According to Da (1984), it makes it possible to gather as much information as possible, store it and organize it in the form of quantifiable, communicable models for a study of the evolution of phenomena and the processes that drive them. It's an approach that makes it possible to establish a spatial differentiation of facts (Hauchart, 2005). This approach is also favoured in biogeographical research, which is the field of investigation of this study.

The systems approach is combined with a hypothetico-deductive approach based on research questions, hypotheses, variables and indicators.

Complex interactions exist between man and the elements of the natural environment (soil, climate, forest, etc.) on the one hand, and the elements of the natural environment between themselves on the other. These interactions result in both direct and indirect influences. This research into the dynamics of plant formations is based on two main logics:

- nature as the main provider of goods and services to the population. Inhabitants of the riparian zones of protected areas derive most of their food from the products of plant formations, medicinal plants are exploited for their therapeutic virtues, and the main source of energy in these environments is wood.

- the second logic is established between man and the natural environment. The relationship between man and nature is complex (Guillard, 2014). They involve spiritual, cultural, economic, political and religious considerations. This millennia-old human intervention in forest resources evolved more or less positively through the creation and conservation of sacred groves; veritable sanctuaries for cultic rites and biodiversity conservation until the end of the 1960s, when population growth, major droughts and the development of the market economy in Sahelian Africa, and particularly in Burkina Faso, triggered an impressive pioneering front that jeopardized these traditional conservation techniques. State intervention to regulate these destructive relationships between man and nature is hampered by the weakness of public institutions in charge of the environment, and by complex cross-cutting issues such as poverty and climate change.

To maintain the ecosystems of Burkina Faso's protected areas in general, and partial and total reserves in particular, it seems undeniable that a balance needs to be struck between removals (logging, clearing for agriculture and livestock farming) and contributions (seeding of local species, reforestation and wildlife farming). Otherwise, reduce the rate of harvesting to the lowest possible level. This will be achieved by implementing participatory management with rural and even urban populations, in compliance with forest legislation.

2.2. Description of study variables and indicators

Several variables are used to test the hypotheses. Table 4 summarizes the hypotheses, variables, indicators and research techniques.

Table4 : Conceptual grid

Main research question: How are plant formations evolving in the Bontioli wildlife reserves? Main hypothesis: plant formations in reserves evolve in fits and starts under the influence of natural and socio-economic factors.

Assumptions	Variables	Indicators	Search techniques
The plant formations on the reserves have a low floristic richness and diversity, and have undergone significant changes between 1986 and 2018.	Plant formations	Growth rate of plant formations	Landsat image processing
		Regression rate of plant formations	
	Floristic richness	Number of species, genera and families	Floristic and dendrometric inventory
	Floristic diversity	Low diversity index	
	Floristic structure	Low stocking densities	
	Vitality of woody plants	Rate of diseased plant species	
The main climate parameters have a negative impact on the evolution of plant formations on reserves	Rainfall	High number of dry years, Annual reduction in waterfall quantities	Literature search Socio-economic surveys Analysis of statistical meteorological data
	Temperature	Average temperature trends Evapotranspiration trends	
	Wind	Wind speed evolution	
	Insulation	Changes in insolation	
Inappropriate farming practices have had a negative impact on the evolution of plant formations in the reserves.	Production systems	Cultivation practices Pastoral practices Grazing on reserves	Literature search Socio-economic surveys Landsat image processing
Local communities are not sufficiently	Involving local communities	Level of involvement	Literature search

involved in reserve management	Payroll organizations	Number of management structures	Socio-economic surveys
	Application of the forestry code	High rate of prohibited acts on reserves	

2.3. Methods

Several methods were used in this study. These included documentary research, socio-economic surveys, meteorological data analysis, floristic inventory, digital processing of satellite images and direct field observations.

2.3.1. Documentary research

Documentary research is the cardinal step in any scientific investigation, as it enables us to take stock of existing documentation. It indicates the current state of information on the topic, and helps to avoid repeating a subject that has already been studied in the same context.

To gather reliable information and draw inspiration from existing studies, we consulted a wide range of documentation, including general works, scientific journals and articles, doctoral theses, master's theses, DEA dissertations, reports from ministries, non-governmental organizations (NGOs), associations, local authorities and websites. For this purpose, documentation centers such as the libraries of the Joseph KI-ZERBO University (the BUC, the libraries of the Geography Departments), of Koudougou, of specialized research institutes (CIRD, Institut Français, CIFOR, IDR), of the Burkina Geographic Institute, of the Ministries of the Environment and Climate Change and decentralized services, of Agriculture, and of the documentation centers of regional and international organizations (CILSS, FAO) have been selected.

Cartographic data were collected for :

- The map of the study sites shows the location of the Bontioli wildlife reserves and the villages involved in the study;

- The morpho-pedological map of the study area shows the distribution of different plant formations according to topography and soil type;

- the map of the reserves' hydrographic network: in principle, watercourses are home to the densest plant communities. This parameter therefore had to be taken into account when describing and characterizing the plant formations;

- the Burkina Faso phytogeographical map to locate Bontioli's wildlife reserves for better vegetation characterization.

Statistical data were also collected at :

-Institut National de la Statistique et de la Démographie (INSD): statistics from the Recensement Général de la Population et de l'Habitat (RGPH) of 1985, 1996 and 2006 for the Bougouriba, Ioba and Poni provinces, the five riverside communes (Zambo, Diébougou, Tiankoura, Nako and Dissin) and the villages bordering the Bontioli reserves. The statistics collected were used to analyze changes in population density.

- Agence Nationale de la Météorologie: rainfall data for the period 1987-2017, as well as temperature, wind and insolation data for the same period from the Gaoua meteorological station were collected.

This station was chosen because it is the main one in the region and the closest to the Bontioli wildlife reserves. Scientific literature also recommends a period of at least 30 years for the measurement of climatic parameters.

2.3.2. Socio-economic surveys

Several surveys were carried out as part of the study.

2.3.2.1. Choice of study sites

Spatially, the research covers the southern Sudanian area of Burkina Faso for the following reasons:

- ✓ This area receives a large flow of agro-pastoral migrants from the north and center of the country;
- ✓ Cotton growing is booming;
- ✓ Artisanal gold-panning sites are scattered throughout the South-West region, particularly in nature reserves;
- ✓ Forests are degrading insidiously.

The Bontioli wildlife reserves were singled out because :

- ✓ This is one of the most important forest complexes in the South-West region, located in the Bougouriba watershed and therefore home to significant biodiversity,
- ✓ Wildlife is still relatively present in these reserves;
- ✓ They have never been developed since their classification;
- ✓ Gold panning is practised in these protected areas;
- ✓ The highly dynamic pioneer front is linked to the ever-growing migratory flow and renewed interest in cash crops, especially cotton, in the riverside villages;

- ✓ The mismatch between public policies, marked on the ground by the occupation of certain parts of the reserves by public infrastructures (social, health and educational).

Villages were selected on the basis of human concentration, proximity of the village to the reserves, the village's economic influence in the commune, i.e. the dynamics of trade (importance of the market in terms of volume of products traded and their diversity) and the importance of agricultural (especially cotton) and pastoral activities. The distance between the villages and the reserves was a determining factor in the choice. Indeed, the human influence on the reserves is all the greater when the dwellings are very close to them, so that the harvesting of forest products can be carried out with little effort on the part of the local populations. Ariori and Ozer (2005) point out that deforestation depends, among other things, on proximity to a timber consumption center and accessibility. The maximum distance between the chosen village center and the reserve boundary is between 0 and 5 km. At this distance, people can easily access the reserves by bicycle or even on foot. Considering the above conditions, thirteen (13) villages were selected for the study.

2.3.2.2. The target population

According to the Petit Larousse de la langue française, a sample is defined as a representative fraction of a certain type of population. In this case, it is a fraction of the population chosen according to the study sites, enabling us to assess the effects of anthropic activities on the evolution of reserves. The method adopted is that of purposive sampling, which consists of determining precise factors in the choice of the target population.

The main users of natural resources (land, vegetation, water, mineral resources) and forest management stakeholders were selected:

- heads of riparian farming households at least 30 years old and having lived in the locality for at least 10 years. These people are expected to witness the changes that have taken place in the reserves. Based on INSD statistics, the number of households per village selected for the surveys was determined according to the formula[14] below:

$n = t^2N/t^2+(2e)^2(N-1)$

n: sample size

N: size of target population (number of households)

t: margin coefficient deduced from the confidence rate, which gives 1.96

e: margin of sampling error ()

The formula is applied to the total number of households in the thirteen (13) villages identified using the RGPH (2006) village , then a proportionality coefficient is assigned to each village according to household size in order to determine the sample of households concerned by the survey. Household heads were selected according to the previously established conditions, i.e. those aged at least 30 and having lived in the locality for at least 10 years.

[14] http://icp.ge.ch/sem/cms-spip/spip.php?article1641 consulted on 14/03/2018

- livestock breeders: this group of stakeholders is essential to the success of any forestry policy, as they are among the main users of natural resources (water and vegetation) in the study area.
- structures involved in forest management (forest supervisors). They are an essential link in the implementation of forest management policies.
- forest service managers. They are the guarantors of the protection, restoration, promotion and implementation of forestry policies and legislation.
- gold miners: their activities are strongly implicated in the regressive dynamics of plant formations and the pollution of water and soil resources.

Table 5 shows the number of heads of household per village surveyed.

Table5 : Target population

Order no.	Villages	Number of households	Proportion	Number of heads of households surveyed	Percentage
1	Bobra-Gogo	133	0,064	21	6,4%
2	Bontioli	205	0,098	32	9,8%
3	Diourao	70	0,034	11	3,4%
4	Djikologo	261	0,125	41	12,5%
5	Forotéon	214	0,102	33	10,2%
6	Gamgbo	88	0,042	14	4,2%
7	Gbonko	152	0,073	24	7,3%
8	Kanséo	46	0,022	7	2,2%
9	Kpankpirè	119	0,057	18	5,7%
10	Manoa	218	0,105	34	10,5%
11	Sinkiro	35	0,017	6	1,7%
12	Zambo	305	0,146	47	14,6%

13	Zoolo	239	0,115	37	11,5%
	TOTAL	2085	1	325	100

Source: INSD, 2006

2.3.2.3. Questionnaire surveys

The theme involves several players in the rural world. These include farmers, stockbreeders and foresters. Nevertheless, a questionnaire is addressed to farmers, who are the main players in the agropastoral production system.

The questionnaire sent out to heads of farming households collected information on the type of farming, the land tenure system, farming practices on the environment and in particular on the forests, the level of knowledge of forestry legislation, compliance with forestry legislation, and the impact of farming and pastoral activities on the reserves.

The questionnaire was first designed on Word 2013, then on SphinxV5. After collection, data were entered on the same SphinxV5 software, sheet by sheet. Statistical graphs and tables were produced using Excel 2013.

2.3.2.4. Surveys using individual semi-directive interview guides and focus groups

The advantage of this type of survey lies in its flexibility. Semi-structured individual interviews were conducted with government officials in charge of the forestry sector, hunterstranshumant herders and gold miners. Focus groups were organized with groups heavily involved in the harvesting and processing of forest products, namely non-timber forest product processors and dolotières .[15]

Discussions with water and forestry officials focused on the status of reserve management, with particular reference to forest law enforcement, the socio-economic uses of forest products and the evolution of plant formations.

During interviews with gold miners, topics covered included mineral exploration, mining and processing techniques, as well as the need to enforce the Forestry Code and restore and protect forest resources.

With the hunters, the aim was to understand the current state of the reserves, hunting practices and their impact on wildlife and plant formations.

Interviews with transhumant herders focused on the use of herbaceous and woody resources for livestock feed, transhumance trails and the application of forestry legislation. Interviews with this group of stakeholders took place in the reserves where they stay. Their positions were first rehearsed by a guide, then the team went out to meet them.

As for the focus groups, they were organized taking into account the following social determinants: gender, age and economic activity. This made it possible to target :

- women who transform sorghum into beer, commonly known as "dolo". This category of economic players is a very important link to be taken into account in the use of the reserves' wood resources. As wood fuel is the main fuel in this flourishing trade in the study area,

[15] Women transforming sorghum into beer (dolo)

we wanted to understand the origin and type of wood (green or dead) used to cook this local beer.

- operators and processors of non-timber forest products: understand the state of availability of these products and the likely reasons for their shortage. Find out how the products are collected. Does it comply with the relevant legislation? Finally, to understand the role of this activity in the local economy. Table 6 shows the various players involved and the types of interviews carried out.

Table6 : Breakdown of players by type of interview

Players	**Survey locations**	**Workforce**	**Type of maintenance**
Foresters	Zambo	01	Individual semi-directive
	Djikologo	03	
	Diébougou	02	
Chaser managers	Bontioli	2	Individual semi-directive
Goldpanners	Zambo	3	Individual semi-directive
	Djikologo	7	
	Zoolo	2	
Dolotières	Zambo	7	Focus groups
	Djikologo	9	
	Zoolo	6	
	Bontioli	7	
NTFP processing group manager	Djikologo	1	Semi-directive
NTFP processors	Zambo	5	Focus groups
	Djikologo	8	
	Bobra-Gogo	7	

Source: Field data, 2020

Qualitative data from individual semi-structured interviews and focus groups have been grouped into sub-themes according to the main ideas developed.

2.3.3. Statistical analysis of meteorological data

Climatic data is a decisive factor in analyzing the evolution of plant formations. To quantify rainfall deficits on different time scales, the Standardized Precipitation Index (SPI) has been determined. This index is used to identify wet or dry periods. It is calculated according to the following formula

ISP= (Xi-Xm) / Si

Xi is the cumulative rainfall for year i

Xm and Si are respectively the mean and standard deviation of annual rainfall observed for the series concerned.

The calculation of this index makes it possible to determine the degree of wetness or dryness of the environment (Bergaoui & Alouini, 2001). In this case, 31 years of data (1987-2017) were used.

When :

ISP>2, humidity is classified as extreme (HE);

1<ISP< 2, high humidity (HF);

0<ISP<1, humidity is moderate (HM);

-1<ISP<0, drought is moderate (SM);

-2<ISP<-1, the drought is severe (SF);

ISP<-2, the drought is classified as extreme (SE).

The umbro-thermal diagram was designed using the formula P=2T with P: monthly precipitation in millimetres and T: temperature in degrees Celsius; this made it possible to identify the start and end of the rainy season (on average).

2.3.4. Floristic inventory

Forest inventory is defined as the action of counting the trees existing on a given surface. It is a method of evaluation that makes it possible to define the information that is essential to any natural resource planning and management policy (Rouamba, 1997).

2.3.4.1. Choice of transects

A partial floristic inventory has been carried out in the Bontioli wildlife reserves. As the overall aim of this survey was to analyze the various floristic parameters, it was important to use a relevant sampling method that would enable the results to be extrapolated to all the plant formations in the reserves. Several sampling methods exist. According to Hadef (2010), there are four methods of investigation:

- systematic sampling: samples are arranged in a repetitive pattern, which may be represented by a network of regular meshes, strips or transects, consecutive segments, grids of points or aligned quadrat points. Survey locations are arranged regularly in space. This type of sampling presents a risk of over- or undersampling of stations.

- random sampling: samples are taken at random from different stations.

- stratified sampling: this method involves dividing a heterogeneous population into more homogeneous strata, then selecting independent samples from each stratum. The sampling method may differ from one stratum to another.

- subjective sampling: based on the observer's perception alone, who judges homogeneity and representativeness.

-Systematic stratified sampling: this method combines the two methods (stratified and systematic) already described. The first step is to divide a heterogeneous population into more homogeneous strata, and the second is to arrange the location of surveys in each stratum in a regular fashion.

The systematic stratified sampling method using transects of variable length was adopted in order to integrate the maximum number of elements involved in the plant formations, with a view to satisfying the criterion of ecological homogeneity, in this case the physiognomy of the flora. Homogeneity also concerns soil type, topography, light, soil humidity, etc. "Within a given perimeter, plant species are distributed according to the influence of different ecological factors (geological nature, substrate, soil, water, relief, altitude, climate, human actions, etc.); consequently, different environments correspond to different floras, and the survey of all environments within the perimeter should enable us to encounter all the taxa present" (Descoings, 1970). This method has a dual advantage in that it is reliable, fast and less expensive (Jiagho & *al.,* 2016). It seems more appropriate for this study in that the aim is to understand the evolution and characteristics of each plant formation.

To do this, the various topographical, pedological and plant units were first identified on the basis of morphopedological and land-use maps (2018). The geographical coordinates of the units in question were taken from the maps and entered into the GPS; guided visits (three) to the reserves were used to identify them.

Following this investigation, three transects (two in the partial reserve and one in the total reserve) of varying lengths were laid out, from the highest to the lowest point of the landform or vice versa, marked by iron bars implanted in both ends of the transects. Geographical coordinates (Latitudes, Longitudes, Altitudes) were then recorded using GPS along the different geomorphological units present on each transect and the plots installed in each plant formation. The field data were then exported to the computer and transect catemas were created using Google Earth (figures 1, 2 and 3).

Figure1 : Catema transect in RTFB (west-east direction)

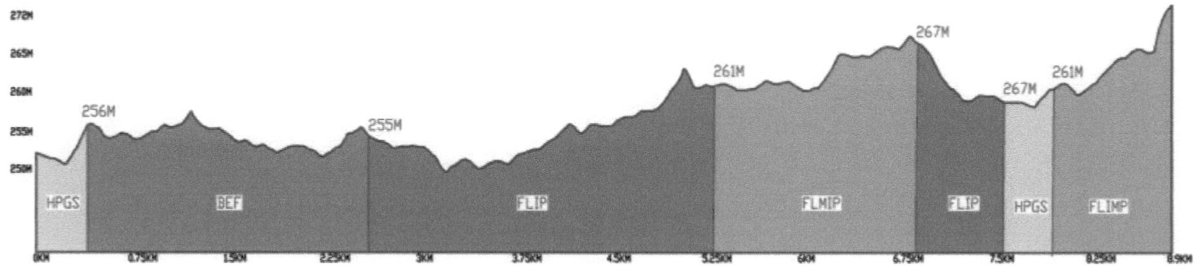

Source: Field data (GPS lerver), December 2020

Figure2 : Catema from transect 1 in the RPFB (south-west-north-east direction)

from

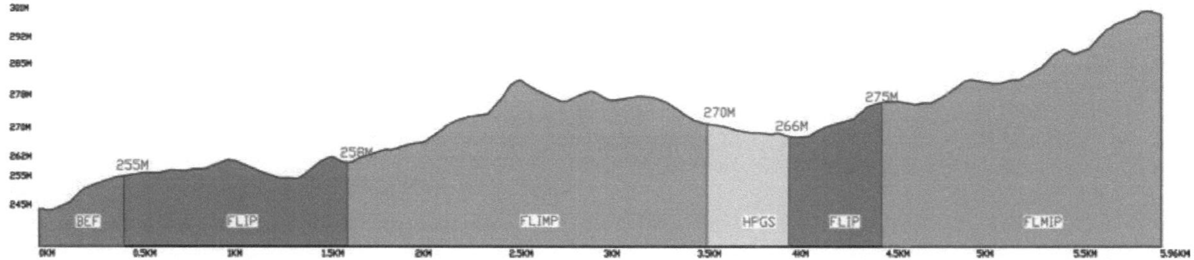

Source: Field data (GPS survey), December 2020

Figure3 : Catema from transect 2 in the RPFB (south-north direction)

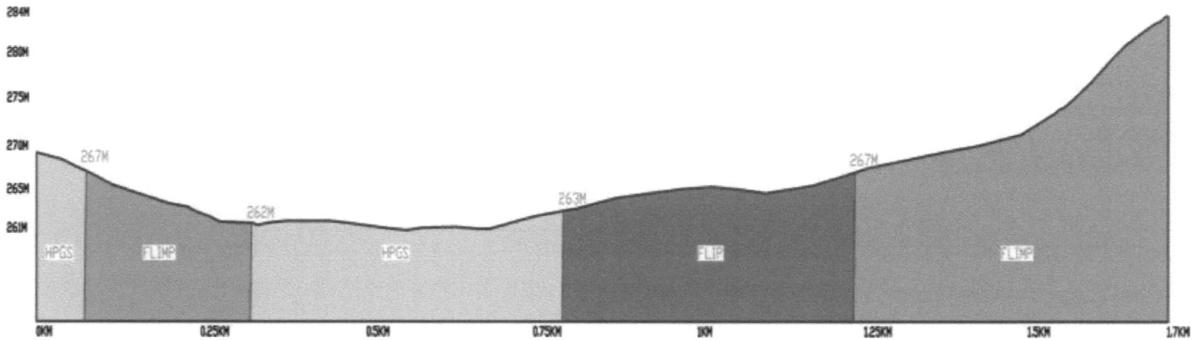

Source: Field data (GPS survey), December 2020

BEF: Ferruginized Tropical Eutrophic Brown soil

FLIP: Ferruginous Tropical Leached Deep Indurated soil

FLIMP: Ferruginous Tropical Leached Indurated Medium Deep soil

HPGS: Hydromorphic to Pseudogley Surface Soil

2.3.4.2. Determining the sampled area, plot selection and equipment

The survey rate is linked to the objectives of the study. The total areas surveyed were 30,000 m² and 75,000 m² for the total and partial reserves respectively, corresponding to 0.024% of the total area of the total reserve and 0.of the total area of the partial reserve. These rates were determined using the formula of Rondeux (1993):

$$F = SE \div ST \times 100\ ;\ \ F \approx F \times \sqrt{1-F}$$

F: polling rate or fraction polled

SE: area sampled (m²)

ST: total surface area (in m²)

The sampling unit chosen is rectangular. According to Jyrki & *al* (1998), rectangular plots allow more species to be recorded than other shapes. The plots are 40 m×25 m (1,000 m²) and 62.5 m×40 m (2,500 m²) in size, and were set up respectively in the densest (gallery forest and wooded and wooded savannahs) and sparsest (shrub savannah, agroforestry park and fallow land) formations. These areas were inspired by the minimum plot area values for the different vegetation types proposed by Thiombiano & *al.* (2016). For savannahs and open forests, the minimum area is 1000 m² for rectangular plots. Many authors have suggested this basic area for these types of vegetation formations, namely savannahs and open forests (Ouédraogo & *al.*, 2006; Bougnounou & *al.*, 2009), but this area can go up to 2,500 m². A total of 54 plots were set up, 21 in the total reserve and 33 in the partial reserve. In order to maintain the ecological homogeneity of the inventory, the plots were placed at an equidistance of 175 metres in each plant formation of the two reservesso that they did not straddle two different plant formations. The following equipment was used:

- a GPS (Global Positioning System) used to record the coordinates of points along transects and plots;
- 400 metres of string and 4 iron bars, each 1.5 metres long, to delimit the plots;
- a decameter for measuring tree circumferences at chest height;
- Floristic survey sheets, pencil, pen, notepad;
- pruning shears to cut stems and leafy branches;
- a bag for collecting tree specimens ;
- newsprint and presses for storing specimens for drying;
- a digital camera for shooting.

The inventory team is made up of six people, including a guide who is a "dozo" self-defense member and village official responsible for protecting the total reserve. Given the security situation and his knowledge of the terrain, his main task was to guide the team.

The five members are divided into three groups: one person is responsible for recording all the information relating to the survey, while the other two groups measure the various plant parameters (botanical names of plant species, girths, heights, state of health) and communicate them aloud. It should be noted that the presence of a forester in the team was of great importance in determining the scientific names of the various plant species during the inventory. The herbarium was not used, as almost all the trees were identified on site. Only two species were photographed for identification purposes. The census concerned adult individuals with a circumference greater than or equal to 30 cm (i.e. a diameter greater than or equal to 9.55 cm) at chest height (1.30 m). Photo 1 shows a tree circumference measurement at chest height.

Photo1 : Circumference measurement session

Field shot (SOME Corentin, December 28, 2020)

Forked trees were inventoried according to the position of the fork. If the fork is at a height of 1.30 m or more, a single tree is counted. If, on the other hand, the fork is less than 1.30 m high, all branches are counted.

Tree heights were measured by estimation. Of the two elements in each sub-group, one whose height is well known stands upright at the foot of the tree, and his teammate is asked to estimate the total height in relation to his height. The procedure involves estimating the number of times the height of the tree is proportional to the height of the element positioned there; this proportion is then multiplied by its height (in meters) to determine the approximate height of the tree.) This procedure has been applied to solve the problem of equipment (lack of clisimeters) for assessing tree heights. The heights of some trees are such that the use of graduated poles was not an option.

2.3.4.3. Processing floristic surveys

Studying the dynamics of plant formations requires an understanding of their horizontal and vertical structures. The data were processed in Excel 2013. Figure 4 shows the distribution of plots according to plant formations in the Bontioli Total Fauna Reserve.

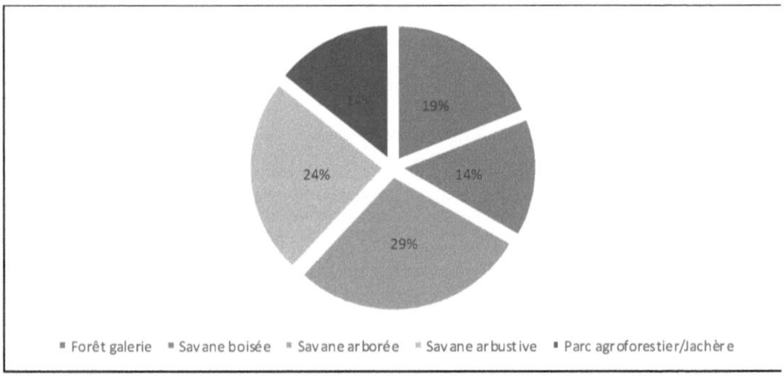

Figure4 : Distribution of plots according to plant formation units in the RPFB

Source: Floristic inventory data, December 2020

Figure 5 summarizes the distribution of plots according to plant formation units in the Bontioli partial wildlife reserve.

Figure5 : Distribution of plots according to plant formation units in the RTFB

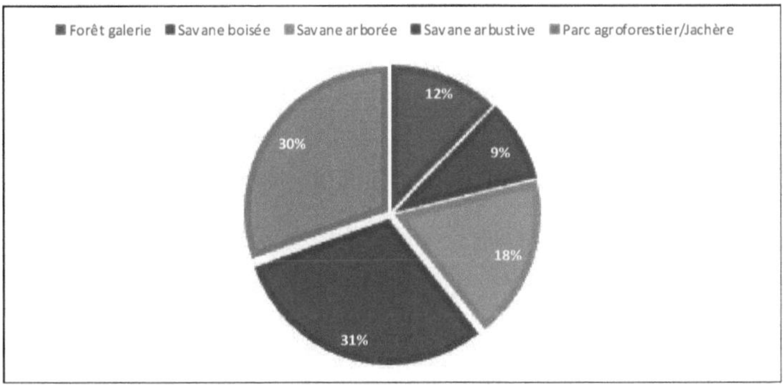

Source: Floristic inventory data, December 2020

According to Cohen (2008), drawing up a list of the species present in the two reserves enabled us to identify the ecological and anthropogenic conditions prevailing there, and to determine the floristic richness. Various indices were then calculated and interpreted:

- specific importance (SI) or relative diversity to determine the most dominant families in the flora. It is the ratio between the number of species in a family and the total number of species, multiplied by 100. It is expressed as follows: IS=Number of species in a family/Total number of species.

- the specific quotient (SQ), which is the ratio of the total number of species identified in a vegetation unit to the total number of genera. It is expressed as QS= S/Ge, with S= the total number of species in the vegetation unit under consideration, and Ge = the number of genera in the same unit. A high value of this index indicates a plant stand rich in species and poor in genera. On the other hand, a low value of this index means that the stand is rich in genera and poor in species. Stability is achieved when the specific quotient is 1 or very close to it. This index is used to characterize vegetation (Lebrun, 1960).

- the Shannon-Weaver Diversity Index (SWDI) to measure the species composition of stands and their relative abundance. It quantifies the heterogeneity of specific diversity (Gnoumou & *al.,* 2021). Its expression is :

$$H' = -\sum_{i=1}^{S} pi \ln pi$$

pi= ni/N, N = total number of individuals, ni = total number of individuals representing species i, pi = relative abundance of species i. The base of the logarithm can be chosen arbitrarily.

This index varies from 0 to lnS ($0 \leq H' \leq \ln S$). It is expressed in bits. It tends towards 0 when the stand is dominated by a single species, while the maximum value is reached when individuals of all species in the stand are evenly distributed.

- woody density (D): this is expressed according to the following formula:

D= N/S

N = number of stems in the plots in the environment under consideration ;

S = total plot area in hectares.

- basal area (G)

$$G = \sum c^2/4\pi \times 1/SE$$

G= basal area in m^2/ha

C^2= circumference at breast height (1.30 m) in m^2 and SE = area sampled in hectare.

- Piélou's equitability index or regularity index :

E= H'/Hmax

H' = Shannon-Weaver index; Hmax= ln(S); S= species richness.

- Piélou's equitability index (E) lies between 0 and 1. If it tends towards 0, this means that almost all the population corresponds to a single species in the stand. When it tends towards 1, each species is represented by the same number of individuals or the same cover. It can be used to assess the weight of each species in the occupation of a plot. This weight may depend, among other factors, on societal choices, climatic conditions favorable to the development of certain species or not, and morphopedological conditions (soil type and topography) of the land.

-Simpson's index (D), which measures the probability that two individuals randomly extracted from the stand belong to the same species. It is a dominance index (Somé & *al.,* 2011) that combines species richness and equitability. Its expression is :

D= Σ $(Pi)^2$. The inverse Simpson index (1-D) was also determined.

All these indices, which do not express the same sensitivity, are used to characterize the abundance and diversity of plant formations in the two reserves. Indeed, the Shannon index is more sensitive to the occurrence of rare species, whereas the Simpson index reflects more the effect of dominant species (Rondeux, 2021).

2.3.5. Digital processing of satellite images

The interpretation of satellite images is a key factor in the analysis of land use and therefore in the study of forest ecosystem dynamics.

Landsat Thematic Mapper (TM), Enhanced Thematic Mapper Plus (ETM+) and Operational Land Imager-Thermal Infrarid Sensor (OLI-TIR) satellite images from 1986, 2002 and 2018 were selected, taking into account the date of recording and image quality, and also with the aim of having a long period to better understand land use changes. The characteristics of the images are shown in Table 7. 1986 is a pivotal date in the evolution of forest management policies, as it corresponds to the year in which participatory management of natural resources, and in particular forests, was implemented in Burkina Faso. One year earlier, in 1985, the government had adopted important policies to protect the environment, namely the "Three Struggles". It seems necessary to assess the impact of this approach at the halfway point (16 years) on land use in 2002. This was the year in which the Land Occupancy Database (BDOT) was renewed. Between 2002 and 2018, 16 years have passed.

Table7 : Satellite image characteristics

Satellites and sensors	Years	Source	Reference	Acquisition date
Landsat 5 (30 m resolution) Thematic Mapper (TM)	1986	Earth Explorer (USGS)/ NASA	LT05_L1TP_196052_19860109_20170218_01_T1 LT05_L1TP_196053_19860109_20170218_01_T1	July 16, 2020, 12 :11 :57

Landsat 7 (30 m resolution) Enhanced Thematic Mapper Plus (ETM+)	2002	Earth Explorer (USGS)/ NASA	LE07_L1TP_196052_20021028_20170129_01_T1 LE07_L1TP_196053_20021028_20170127_01_T1	July 17, 2020, 11 :54 :12
Landsat 8 (30 m resolution) Operational Land Image (OLI) / Thermal Infrared Sensor (TIRS)	2018	Earth Explorer (USGS)/ NASA	LC08_L1TP_196052_20180306_20180319_01_T1 LC08_L1TP_196053_20180306_20180319_01_T1	August 5, 2020, 12 :39 :20

The various image processing stages are: pre-processing, classification, post-processing (image smoothing and homogenization), vectorization and transition matrix determination.

2.3.5.1. Satellite image pre-processing

It is essential to pre-process satellite images before processing them. The aim is to increase data readability and facilitate better information extraction. In practice, this stage consisted of radiometric correction, including data correction due to sensor irregularities, sensor or atmospheric noise, and geometric correction due to variations in Earth-sensor geometry. Images were enhanced to improve appearance and facilitate visual interpretation and analysis. All preliminary processing was carried out using ENVI 5.3 software.

2.3.5.1.1. Image enhancement

Pratt (1991) defined image enhancement as a set of techniques that seek to improve the visual appearance of an image, or to convert the image into a form more suitable for human or machine analysis. For image enhancement, it was necessary to :

- scale amplitude and modify contrast ;
- modify the histogram to subdivide it into non-adaptive and adaptive modifications, then linear and non-linear clean-ups;
- Contour enhancement with linear enhancement and statistical differentiation ;
- color image enhancement, including natural color, pseudocolor and false color image enhancement;
- enhance the multi-spectral image.

2.3.5.1.2. Radiometric correction

In visible and near-infrared passive remote sensing, radiation is emitted by the sun, passes through the atmosphere and arrives at the earth's surface. This radiation is then recorded by the satellite sensor. Modifications to the original radiation occur along the way, due to the effects of the atmosphere and the earth's own elements. Radiometric enhancement consists in correcting the effects of the various artifacts that disrupt radiometric measurement, notably sensor defects and atmospheric haze. It involves redistributing the gray levels (or colors) so as to make better use of the available palette. This is known as histogram *stretching*.

2.3.5.1.3. Geometric correction

Remote sensing images need to be corrected for spatial distortions in order to maximize their usefulness for information extraction and later integration with other information sources, such as a GIS database. Geometric correction involves repositioning image pixels in their normal positions. To achieve this, the pixel resampling procedure was applied in this study. Resampling has the advantage of ensuring that pixel supperpositions and comparisons are correct.

2.3.5.1.4. Color composition, scene mosaics and study area extraction

Processing of the satellite images began with the creation of a color composition using the bands concerned, a mosaic of the scenes, extraction of the study area and collection of the coordinates of certain points on the image (color composition), which were used to create vector layers of the training sites. To facilitate the discrimination of land use units, a combination of three (03) bands was made using the three primary colors (red, green and blue). These were mid-infrared, near-infrared/red (bands 6/5/4) for Landsat 8 OLI-TIR scenes and 5/4/3 for Landsat 7 scenes. The 6/5/4 color composition has been used by several authors in the analysis of land cover and forest ecosystems (Tiesse & *al.,* 2017; Cabala Kaleba & *al.,* 2017; Diallo & *al.,* 2020). And mid-infrared, near-infrared/red (4/3/2 bands) for ETM+ and TM scenes. These compositions were chosen because they better discriminate vegetation from other occupancy units.

In addition, the coloured composition is read by looking at it and deducing the correspondence of the vegetation or land-use classes shown on it (Table 8), based on knowledge and field data.

Table8 : Key to interpreting satellite images

Units	Interpretation color
Water level	Blue
Gallery forest	Dark green
Wooded savannah	Green
Shrubby savannah	Light green

Rainfed cultivation	White green
Bare ground	White

Scene mosaicing was carried out because the study area straddles two (02) scenes (196-052 and 196-053), i.e. the scenes had to be combined before the study area could be extracted. Mosaicking is therefore necessary. The study area has been cut across the entire mosaicked image.

The training site vector layers were created by taking into account all the colors present on each color-composed image of the study area. In fact, the most recent image (2018) in color composition was explored and all forms of land use were identified. Detecting the various land-use units from satellite images alone remains difficult. For this reason, it was necessary to rely on data collected in the field. On the one hand, this enabled us to check the exact coordinates taken from the image (training sites), and on the other, to observe and survey the GPS coordinates of certain specific land-use units, which were used to check their colors on the color-composition images. Once this had been done, the creation of training site vector layers was continued. The training site vector layers of the 1986 and 2002 images were thus inspired by those of the 2018 image in terms of identifying land-use units according to the different colors.

2.3.5.2. Image processing proper

This step is subdivided into three sub-steps.

2.3.5.2.1. Supervised classification and the Kappa index

The satellite images were then processed in three main phases: supervised classification, post-classification and vectorization.

The types of classification are very diverse but can be divided into two categories: supervised classification, which uses a training dataset to calibrate the algorithm beforehand, and unsupervised classification, which produces clusters of pixels that will be labeled a posteriori as a ground cover class based on in situ data or auxiliary information (Defourny, 2017).

Classification is based on the radiometric value of the occupancy units to be classified. Classification is supervised or directed, and the method is based on the Maximum Likelihood algorithm, which is characterized by a distribution of pixels by class. The pixel is assigned to the class with the highest probability. This method classifies all pixels unless a probability threshold is applied (below which pixels with very low probability will not be classified) (Ouédraogo, 2015). After this classification, the confusion matrix was generated. Classification is evaluated by this matrix and the Kappa index. The confusion matrix identifies errors of omission at the level of each land-use unit during interpretation and other errors due to confusions between land-use units (Yanogo, 2013). This matrix is used to calculate the Kappa coefficient or Kappa index (total number of pixels well classified in its class/total number of pixels in the class in the field). This is an indicator of the quality of the classification. It lies between 0 and 1. When its value tends towards 1, images are relatively well classified. On the other hand, classification is unsatisfactory if this index tends towards 0. The results obtained are shown in Table 9.

Table9 : Overall accuracy and Kappa index

Years	Overall accuracy	Kappa index
1986	300/321	93,45%
2002	278/371	74,93%
2018	500/543	92,08%

Source: Satellite images from 1986, 2002 and 2018

The relatively high accuracies resulting from the classification of the different images are linked to the reduction in the number of thematic classes (5). Wooded savannah was included in treed savannah because these two plant formations have the same reflectance. They also have the same vegetation cover (30-60%). Open forest has almost disappeared from the area, and in places is similar to gallery forest.

2.3.5.2.2. Post classification and vectorization

For this operation, the classes were first combined using the "combine classe" tool; isolated pixels were then eliminated using the "seive" tool. In addition, "clump classe" was used to homogenize the classes and "Majority/Minority analysis" to smooth them.

After post-classification, the classified image was vectorized and transferred to ArcGis10.3 software for cartographic processing and analysis. The habitat will be further digitized using ArcGis10.3 software.

The transition matrix is created using Excel 2013. It is a dynamic double-entry table that allows analysis of the various changes in land-use classes over the course of two distinct dates. The transition matrix provides a condensed description, in the form of a square matrix, of land-use changes and allocations over the time interval under consideration (Scholaepfer, 2003 cited by Ouoba, 2013).

In the matrix, the rows represent the land-use units at time t0 and the columns the land-use units at time t1. Area mutations take place from a class i at time t0 corresponding to the initial area to a class j representing the final area at time t1. The total area of the study area corresponds to $\Sigma\Sigma\, a(i,j)$, while $Eit0 = \Sigma\, a(i,j)$ is the sum of line i corresponding to the total area of land-use unit i at time t0. The sum $Ejt1 = \Sigma\, a(i,j)$ of column j represents the total area of land-use unit j at time t1 (Yanogo, 2013). Elements outside the diagonal represent changes in land use (Bamba, 2010). Table 10 shows an example of a transition matrix between two dates t0 and t1.

Table10 : Example of a transition matrix between dates t0 and t1

		Land use units j at time t1			
		Unit 1 (j=1)	Unit 2 (j=2)	Unit 3 (j=3)	Sum of Eit0 lines
Land-use units i at time t0	Unit 1(i=1)	a(1,1)	a(1,2)	a(1,3)	$E1t0 = \Sigma\, a(1,j)$
	Unit 2 (i=2)	a(2,1)	a(2,2)	a(2,3)	$E2t0 = \Sigma\, a(2,j)$
	Unit 3 (i=3)	a(3,1)	a(3,2)	a(3,3)	$E3t0 = \Sigma\, a(3,j)$
	Sum of Ejt1 columns	$E1t1 = \Sigma\, a(i,1)$	$E2t1 = \Sigma\, a(i,2)$	$E3t1 = \Sigma\, a(i,3)$	$\Sigma\Sigma\, a(i,j)$

The transition matrix is characterized by three main phases:

- the stability phase corresponds to the areas on the diagonal ;
- the gain phase, whose surfaces can be seen below the diagonal;
- the loss phase, with areas at the top of the diagonal.

Figure 6 summarizes the various satellite image processing stages.

Figure6 : Summary of the main satellite image processing stages

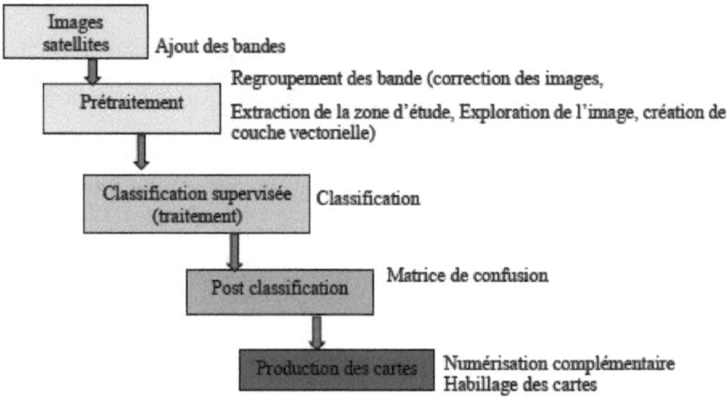

2.3.6. Direct observations in the field

The observations consisted of a field visit to the two reserves, independently of the floristic inventory and other field work. The aim was to identify slash and burns associated with the presence of man and his activities in the reserves, as well as the influence of climatic parameters on the environment. An observation sheet has been drawn up for this purpose. It takes the form of a table containing the following headings: slash, period of visit and observations. The observations identified the following elements:

- agricultural activities within the reserves were observed (fields, new clearings and the presence of exotic species, in this case *Tectona grandis*);

- pastoral activities (presence of domestic animals, their droppings, trampling of shrubs, soil compaction around the Mouhoun and Bougouriba rivers by herds);

- other human activities (bush fires, pruned or completely cut trees, debarked trees, presence of dwellings or house ruins in reserves, gold panning holes, etc.).

The effects of natural factors on vegetation (trees uprooted by wind, dead trees caused by drying, etc.).

- some actions of the Forest Investment Program (PIF) in the reserves, i.e. an area set aside in the total reserve, surrounded by fencing to prevent herd intrusions, has been observed.

The field visit was an opportunity to test the data collection tools, i.e. the questionnaire and interview guides, in order to identify and correct any shortcomings prior to their administration.

Systemic and hypothetico-deductive approaches were chosen for this research. With this in mind, various tools and techniques were used: qualitative studies (guides for semi-directive interviews and focus groups, field observation), quantitative studies (survey questionnaire), floristic study techniques, analysis of climatic parameters and land use. Methods for processing statistical data collected in the field, i.e. meteorological data, satellite images and data from the floristic inventory, were presented.

Partial conclusion

This part of the thesis, devoted to the theoretical framework and methodological approach, was used to justify the study and formulate the hypotheses, objectives and variables. An analysis of the state of knowledge on the subject revealed that, overall, a wide range of factors influence the dynamics of plant formations. Nevertheless, anthropogenic factors, in this case agriculture, livestock farming and forestry, and natural parameters, in particular climatic variability, are highlighted in the scientific literature. With a view to orienting the study within a well-defined epistemological framework, theories relating to demography and the evolution of natural resources and biogeography have been set out and analyzed.

Data collection, processing, analysis and interpretation were based on quantitative, qualitative, digital satellite image analysis and mapping methods.

PART TWO: THE INFLUENCE OF THE BIOPHYSICAL AND HUMAN ENVIRONMENT ON PLANT FORMATIONS

This section is divided into three chapters:

- ✓ Chapter 3: The biophysical and human environment. It deals with the physical characteristics and political and religious organization of the local populations, as well as demographic dynamics.

- ✓ Chapter 4: Land use and characterization of reserve plant formations. The state of land use and the evolutionary dynamics of plant formations in 1986, 2002 and 2018 were reviewed. Floristic richness and diversity and the structural parameters of plant formations were also analyzed.

- ✓ Chapter 5: The effects of climatic sustainability on reserve plant formations. Changes in the main climatic parameters, i.e. rainfall, ETP, temperature, wind and insolation over the 1987-2017 series, in relation to changes in plant formations, were deciphered.

CHAPTER 3: THE BIOPHYSICAL AND HUMAN ENVIRONMENT

This chapter presents the biophysical characteristics of the Bontioli wildlife reserves, i.e. their vegetation, morphological and edaphic aspects. On a human level, the analysis of demographic dynamics and socio-political organization has enabled us to address the activities carried out by local populations.

3.1. Geographical location and physical characteristics of the study area

In this section, we define the exact location of the study area by examining its physical characteristics.

3.1.1 Geographical location of the Sud-Ouest region

The South-West administrative region was created by law N°2001-013/AN of July 2, 2001 creating the regions. It comprises four provinces: Bougouriba with Diébougou as its capital, Ioba with Dano as its capital, Noumbiel with Batié as its capital and Poni with Gaoua as its capital. It is bordered to the east by the Republic of Ghana and the Centre-Ouest region, to the west by the Cascades region, to the south by the Republic of Côte d'Ivoire, and to the north by the Hauts-Bassins and Boucle du Mouhoun regions. It covers of the national territory, or 16,318 km². Due to its geographical position, south of the Sudanian domain, the South-West region offers significant agricultural potential, in the form of lowlands that can be developed, forestry potential such as classified forests, reserves and protected forests, fishing potential (water reservoirs and natural watercourses) and pastoral potential (fodder for livestock).

This region is relatively less populated, with a total population of 620,767 in 2006, the majority of whom live in rural areas (550,553 inhabitants) (RGPH, 2006) and therefore carry out rural activities, the main one being agriculture. According to the results of the 2006 RGPH, the region's average population density (38.0 inhabitants/km^2) remains below the national average of 51.8 inhabitants/km^2. The same applies to the growth rate of 2., compared with the national rate of 3. at the same date. Projections for 2015 indicate slightly higher rates than the indices listed.

Several ethnic groups live together in this geographical and administrative area. But the most representative in numerical terms are the Lobi, Dagara, Birifor, Djan, Gan, Pougli and Mossi.

3.1.2 Location of study sites

The Bontioli total and partial wildlife reserves, established respectively, by decree no. 3147/SE/EF of March 23, 1957 delimiting and classifying the total wildlife reserve (RTFB) covering an area of 12,700 ha, and by decree no. 3417/SE/EF of March 29, 1957 delimiting and classifying the partial wildlife reserve (RPFB) covering an area of 29,500 ha, are located between latitudes 10°40' and 10°56'N and longitudes 2°53' and 3°09'W in the South-West region (see map 1). Covering a total area of 42,200 ha, they straddle the provinces of Ioba, Bougouriba and Poni, and extend over the communes of Zambo, Diébougou, Tiankoura, Nako and Dissin. They are located on more than 90% of the surface area of the commune of Zambo (MEEVCC, 2017).

Map1 : Area and location of study sites

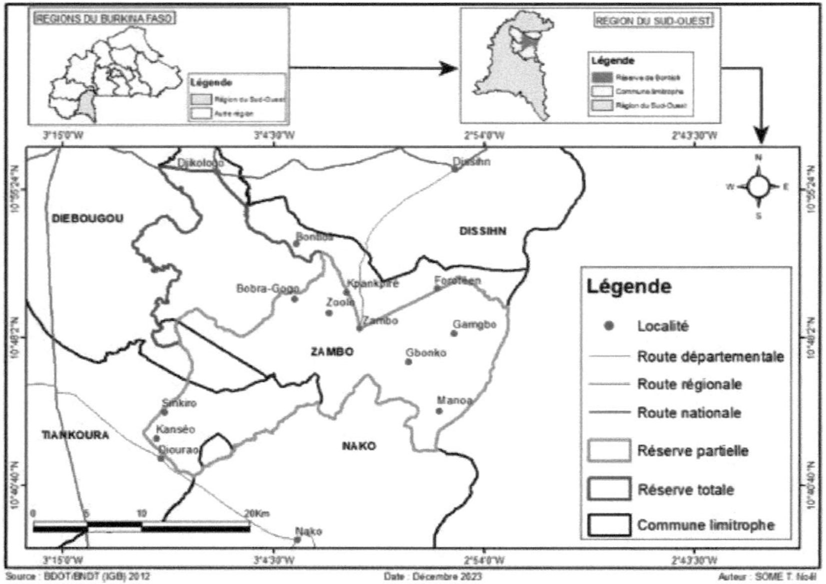

3.2. The biophysical environment

Knowledge of the biophysical environment is essential for analyzing the dynamics of plant formations.

3.2.1. Morphopedological and hydrographic characteristics of the reserves

3.2.1.1. Relief

The Bontioli wildlife reserve area is very flat. Average altitudes are between 200 and 350 m for the total reserve and 200 and 450 m for the partial reserve. Despite this generally flat relief, the area is dotted with hills, armoured buttes, shallow valleys and gullies. Given the flatness of the overall relief, the water in these valleys runs close to the ground. They dry out rapidly in the dry season and have a sinuous appearance. Alluvial terraces line the riverbanks.

3.2.1.2. Morphopedological units

Soils in the Bontioli wildlife reserves are related to topography, underlying geological formation and climate (Map 2). Based on the classification of soils carried out by the CPCS (1967), an outline of soil typology in relation to the topography of the area has been drawn up:

- lithosols on rock: these cover 0.of the reserve area (71.16 ha) and are found on rocky knolls. These are shallow soils with little or no agronomic value.
- lithosols on armour are found on armour-clad hills and plateaus. They cover 2.37% of the reserves' surface area (1 098.63 ha). Like the preceding soils, they are shallow and of low agronomic value. These soils are unsuitable for cultivation.

On these two types of soil, a shrub savannah-type plant formation has essentially developed, due to the importance of rocky outcrops. The plant species are stunted shrubs.

- surface ferruginous leached indurated soils are very little represented on a total surface area of 8.75 ha, i.e. 0.. They are located on plateaus and have low to medium agronomic suitability.

- medium-deep leached indurated ferruginous soils cover 14,790.7 ha, or 31.91% of the reserve area. This is one of the most important soil types in terms of surface area. They have average agronomic value. They form on medium-slope glacis, upper glacis slopes and plateaus.

- deep indurated ferruginous leached soils: these occupy the lower slopes of the glacis and the middle slopes of the glacis, covering an area of 20,129.1 ha or 43.of the area of the reserves. They have average to good agronomic value.

These two types of soil are used for agriculture because of their high agronomic value. On these soils, dense plant formations of the tree savannah type have developed. The density of woody plants increases on deeper, better-drained soils (Fournier, 1991).

- ferruginized tropical eutrophic brown soils: these cover 2,877.47 (6.) and are found on alluvial terraces. They have good agronomic value.

- Vertic brown eutrophic tropical hydromorphic soils and brown eutrophic tropical hydromorphic soils: these are the least represented in the reserves, with a very small surface area compared with other soil types (3.69 ha or 0.01%). They cover the lower slopes of the glacis.

- hydromorphic soils of alluvial origin: these are identified on the banks of watercourses. They cover 79.26 ha, i.e. a total reserve area of 0..

- low-humus hydromorphic soils with pseudo gley surface: located in lowlands and flood-prone areas. These soils cover a surface area of 7,075 ha, or 15.of the reserve area. Part of the gallery forest has settled here, with hygrophilous plant species such as *Mitragyna inermis*. These soils have a high agronomic value and are used for rice production and market gardening during flood recession periods.

Map2 : Morphopedological units of the Bontioli wildlife reserves

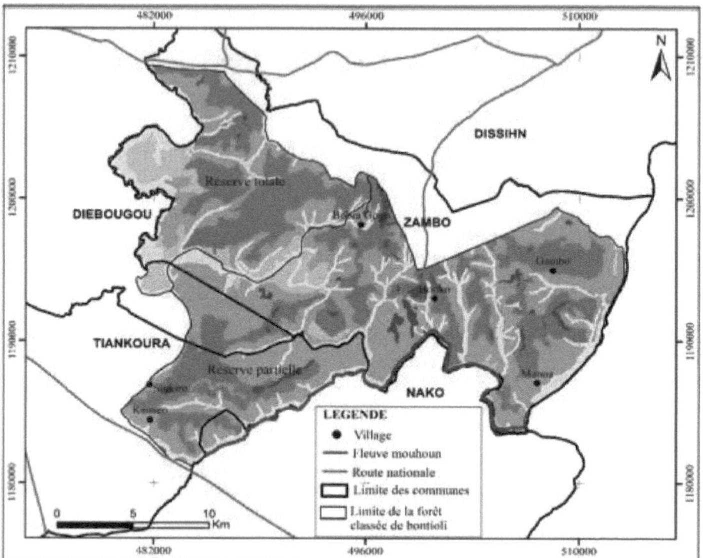

Unité	Type de sol	Géomorphologie	Superficie	
			Ha	%
	Litbosols sur roche	Butte rocheuse	71,16	0,15
	Litbosols sur cuirasse	-Butte cuirassée -Plateau cuirassé	1098,63	2,37
	Ferrigineux tropicaux lessivés indurés superficiels	Plateau	8,75	0,02
	Ferrigineux tropicaux lessivés indurés moyennement profond	-Glacis pente moyenne -Haut de pente du glacis -Plateau	14790,7	31,91
	Ferrigineux tropicaux lessivés indurés profond	-Bas de pente de glacis -Moyenne pente de glacis	20129,1	43,43
	Bruns eutrophes tropicaux ferruginisés	Terasse alluviale	2877,47	6,21
	-Bruns eutrophes tropicaux hydromorphes vertiques -Bruns eutrophes tropicaux hydromorphes	Glacis bas de pente	3,69	0,01
	Peu évolués d'apport alluvial, hydromorphes	Bourrelet de berge	79,26	0,17
	Hydromorphes peu humifères à pseudo gley de surface	-Bas-fond -Zone inondable	7075	15,27
	Plan d'eau	-	210,5	0,45
			46344,26	100

Source: BUNASOLS; Mars 2020

3.2.1.3 The hydrographic network

The Bontioli wildlife reserves benefit from an abundant hydrographic network (see map 3). They are drained by two major rivers, the Mouhoun and its tributary, the Bougouriba. They are located in the

Bougouriba watershed. The Mouhoun, Burkina Faso's main river, forms the natural boundary of the partial reserve with Ghana, while the Bougouriba drains its course between the two reserves. In addition to these main watercourses, numerous marigots and rivers are scattered in and around the reserves. Apart from the Mouhoun, which has a permanent regime, the others are temporary streams that dry up totally or partially, creating a string of pools (Bougouriba) in the dry season. This creates a migration of pachyderms towards the Mouhoun.

This hydrographic network is of cardinal importance for the reserves and the villages along their banks. The availability of water in these rivers, especially the Mouhoun, helps to maintain wildlife in the reserves, and even attracts certain migratory species such as birds. These water reservoirs are also an important resource for agricultural (market gardening), pastoral and fish farming production in riverside villages. Finally, they have a positive effect on plant formations, as they shelter gallery forests.

Map3 : Hydrographic network of Bontioli wildlife reserves

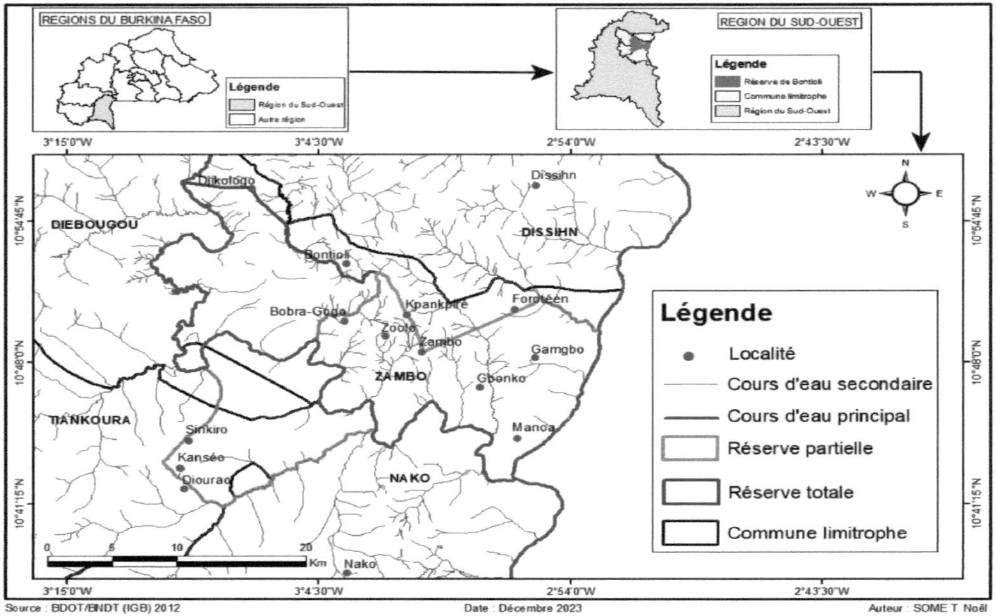

3.2.2. The climate

Climate is an essential factor in the spatial distribution and growth of plants in tropical zones.

3.2.2.1. The climate mechanism in Burkina Faso

The West African climate, and by extension that of Burkina Faso, is influenced by prevailing winds originating from three high-pressure areas: the St. Helena High, the Saharan High and the Azores High. These winds converge in a zone called the Intertropical Convergence Zone (ITCZ) or Intertropical Convergence (ITC). The surface position of the ITCZ is located in the low-pressure or discontinuity zones of converging winds[16] . It is a crossing zone for the maritime trade winds blowing in a south-west-north-easterly direction: the monsoon. It originates from the St. Helena anticyclone in the Atlantic Ocean, positioned around 10°W. As it moves northwards, it becomes warmer and wetter (Isnard, 1952), and the harmattan, a dry continental trade wind originating from the Saharan depression south of Libya, picks up sand and dust as it crosses the desert towards the Gulf of Guinea. It sucks in the cold, damp trade winds from the Azores anticyclone, which is positioned over the Atlantic Ocean at 35°N in January and 40°W in July, while at altitude it is at 15°N in January and 27°N in July (Vissin, 2007). It's a dry wind, warm during the day and cold at night, blowing in a northeast-southwest direction. These two winds, circulating in the lower tropospheric layers, determine the rainfall mechanism in West Africa. It generates the majority of rainfall in the Sahel, which results

[16] Meteo-area.blogspot.com consulted on 2/5/2021.

from isolated convective thunderstorms organized in squall lines (Hountondji, 2008). Vischel, T. & *al.* (2015) note that rainfall in the Sahel is produced by large-scale thunderstorm systems propagating from east to west at tens of kilometers per hour, known as mesoscale convective systems (MCS). Its latitudinal position varies with the seasons, depending on the declination of the sun. The ITCZ, over the continent, is positioned around 10°N in July, where the energy available for convection is at its highest. In August, the ITCZ reaches its northernmost position over the continent at around 19°N in boreal summer (July/August).

The area is covered by heavy rainfall. In September, it begins a reverse migration to winter over the ocean near the equator (Karam, 2008). These movements are associated with the north/south movement of the East African jet over West Africa (Ouoba, 2013). It is a strong easterly wind whose average intensity varies between 10-15 $m.s^{-1}$ in summer and which extends between 30°E and 30°W and its latitudinal position varies between 12 and 15°N between June and September (Poan, 2013).

The alternating northward and southward movements of the intertropical convergence determine the orientation of climatic zones and vegetation zones in West Africa according to latitude bands (Dykstra & *al.,* 1996).

Given its southern location, the South-West region is one of Burkina Faso's wettest. It benefits from a south-Sudanese climate, limited to the north by the 900 mm isohyet.

3.2.2.2. Burkina Faso's climatic zones

Burkina Faso is subdivided into three major climatic zones based on rainfall in space and time:

- the Sahelian zone, limited to the south at 14°N and by the 600 mm isohyet. The rainy season lasts only about three months (June to September). Temperatures and thermal amplitudes are high. This is the domain of shrub steppes.

-North Sudan: located between 11°30' and 14°N, with isohyets of 600 mm and 900 mm to the north and south respectively, the winter season extends from June to October, lasting around five (5) months. Temperatures are relatively high, especially in the northern part of the region, as are thermal amplitudes. The vegetation consists of an association of Sahelian-type plant formations in the north (shrub steppe) and Sudanian-type formations in the center and south (shrub savannah, grassland, trees and woodland).

- South Sudan: located south of the 11°30' parallel and bounded to the north by the 900 mm isohyet. This is the country's wettest area. The rainy season lasts between 6 and 7 months, from April-May to October. Temperatures are lower than in the two above-mentioned areas. Daily and monthly temperature ranges are relatively low. Vegetation consists of a mosaic of savannahs (grassy, shrubby, wooded, wooded) and forest galleries on the edges of watercourses.

3.2.3. The vegetation

The Bontioli wildlife reserves are located in the southern Sudanian climate, precisely in the southern Sudanian phytogeographic domain, western Volta-Noire (Mouhoun) district (Guinko, 1984), where vegetation benefits from fairly abundant rainfall. The average rainfall for the series studied (1987-2017) based on data from the Agence Nationale de la Météorologie (Gaoua station) is 1081 mm.

The plant cover is made up of a mosaic of plant formations consisting of agroforestry parks with *Vitellaria paradoxa*, *Parkia biglobosa*, *Faidherbia albida* and *Adansonia digitata,* shrub savannahs dominated *by Acacia dudgeoni, Acacia senegal, Acacia seyal, Acacia ehrenbergiana, Acacia laeta, Acacia goumaensis, Combretum glutinosum, Combretum molle, Combretum fragrans, Lannea microcarpa and Detarium microcapum,* wooded savannahs, tree savannahs and gallery forests with *Anogeissus leiocarpus, Diospyros mespiliformis, Mitragyna inermis, Daniellia oliveri* and *Balanites aegyptiaca* along the river and other watercourses. The physiognomic units of the plant cover are arranged according to topography and soil type. Shrub savannah is generally found on lithosols located on plateaus and armoured mounds, which are not very rich in nutritive elements, whereas wooded savannah and gallery forest are found in depressions and riverbanks, where the soil is richer and retains moisture over a long period. A diversity of plant species can be found here, which will be analyzed in the section presenting the results of the floristic inventory. Map 4 shows the phytogeographical zones of Burkina Faso.

Map4 : Phytogeographical regions of Burkina Faso

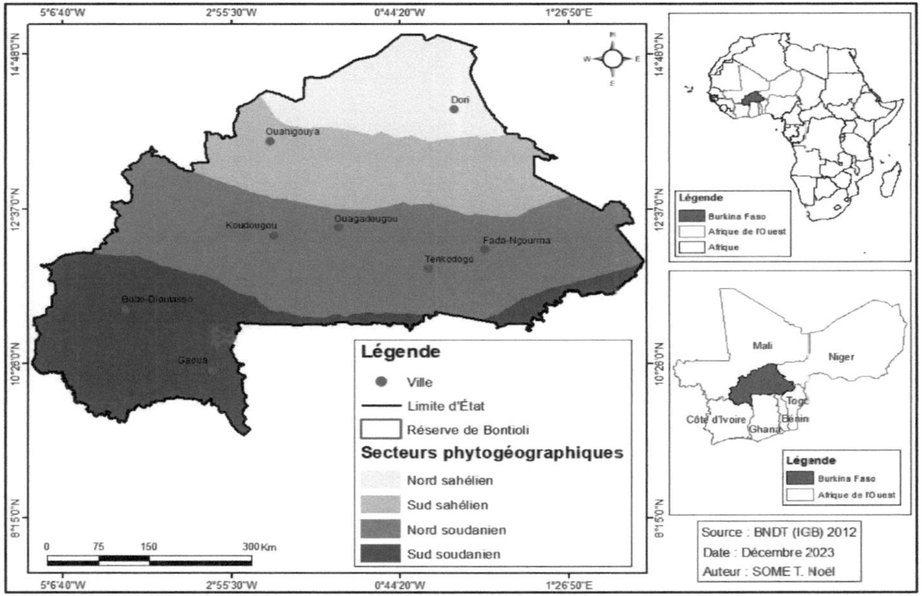

3.3. Human environment

At this point, it would seem advisable to analyze the demographic dynamics and social organization in place, with a view to better understanding their effects on the evolution of forest resources.

3.3.1. Demographic dynamics and social organization

Analysis of demographic dynamics and human activities is essential to understanding the current and future evolution of reserve landscapes. The interaction between the two components of the environment (man and vegetation) remains inseparable. Man's actions influence the environment. Indeed, man shapes the environment according to his way of life, his cultural and spiritual aspirations - in short, his conception of the world. On the other hand, nature offers a wide range of development opportunities for man to choose from, depending on the technology at his disposal and his ability to adapt and innovate. With this in mind, one approach to geography sees space as a multitude of possibilities available to man. It should be pointed out that a company's choice is a decisive factor in the conservation of the resource on which it depends.

3.3.1.1. Population trends in provinces hosting reserves

The general population censuses of 1985, 1996 and 2006 show changes in population numbers and densities in the provinces hosting the Bontioli wildlife reserves. Population densities rose from 23.1 inhabitants/km² for Bougouriba in 1985 to 27.1 inhabitants/km² in 1996 and then to 36 inhabitants per km² in 2006, representing a growth of 4.1 inhabitants/km² in 11 years. Between 1985 and 2006, population density in Bougouriba province increased by 10.9 inhabitants/km². The Ioba province is the most densely populated in the South-West region.

Population density rose from 43.1 inhabitants/km² to 49.7 inhabitants/km² in 1996, then to 59.2 inhabitants/km² in 2006. This represents an increase of 6.6 inhabitants/km² between 1985 and 1996, or 11 years. Over the decade 1996-2006, the population of Ioba province grew even more. The growth rate was 9.5 inhabitants/km². Finally, the Poni province also saw an increase in population density. Statistics show increases of 2.2 inhabitants/km² and 8.4 inhabitants/km² respectively between the periods 1985 - 1996 and 1996 - 2006 (table 11).

Taking into account the 2015 INSD projection, population trends in the three provinces bordering the reserves are quite significant.

Overall, population densities in the region's various provinces increased significantly between 1985 and 2006. This growth was 12.9 inhabitants/km², 16.1 inhabitants/km² and 8.9 inhabitants/km² respectively for the Bougouriba, Ioba and Poni provinces. Compared with the national average density of 51 inhabitants/km² in 2006, and with the exception of the Ioba province, the other provinces in the region were still sparsely populated.

Table11 : Population trends in the provinces of Bougouriba, Ioba and Poni in 1985, 1996, 2006 and 2015

	Populations				Densities			
Years	1985	1996	2006	2015	1985	1996	2006	2015
Bougouriba	65093	76498	101479	131748	23,1	27,2	36	46,8

Ioba	1402 50	1614 84	1923 21	2398 01	43, 1	49, 7	59, 2	73,7 6
Poni	1917 54	1959 00	2569 31	3321 69	26, 1	26, 6	35	45,1 9

Sources: INSD, general population censuses 1985, 1996 and 2006, 2015 projection

3.3.1.2. Population trends in riverside villages

According to the results of the 2006 RGPH, the total population of the thirty-one (31) villages bordering the reserves was 21,387 (INSD, 2006). This population is expected to grow substantially by 2020, according to the Institut National de la Statistique et de la Démographie (INSD), which estimates a population of 30,730, an increase of 43.in 14 years.

Population growth is linked to agricultural, pastoral and gold mining migration, and above all to the local birth rate, which remains high. In the sites surveyed, the average number of children per head of household surveyed was 6. However, 87.4% of respondents had fewer than 10 children, and 10.had between 10 and 20 children. The growth in the area's notified population implies an increase in the need for firewood, new farmland, grazing land and so on. Agricultural and pastoral production systems therefore need to be taken into account in understanding the impact of population growth on natural resources in general and plant formations in particular. Household size, which determines the number of mouths to feed, should have a positive effect on NTFP exploitation and consequently on NTFP dependency (Ouédraogo & *al.,* 2013), since in rural areas, households that rely on forest products are generally the most vulnerable. All these needs are met at the expense of forest areas (Achard & Blasco, 1990). The satisfaction of these needs also obeys the rules inherent in the socio-political and religious organization of the societies living there.

3.3.2. Socio-political and religious organization

The people who settled in the south-western part of Burkina Faso came from Ghana in successive waves between the eighteenth and first half of the nineteenth centuries, and established a socio-political organization that was not centralized, but was very much in evidence at family level. Among the Dagara, in particular, this organization is based on a lineage system. The unit of reference is the large family, made up of relatives descended from the same ancestor who founded the patrilineal lineage. Descendants of the lineage bear the same patronymic (Somé, 2015). The principle of double parental descent is thus applied. The head of the family, also called "yirsob" in the Dagara language, is the head of the household. He is responsible for all the members in his care. It is within the family that the clan's continuity is ensured through marriages that follow the norms prescribed by society. Unions are generally endogamic between individuals from different families, the aim being to safeguard the cultural values of the society (language, cultural practices) and the ethnic group. Unions between members of the same clan or extended family are forbidden; they are considered incestuous. Polygamy is the matrimonial regime most commonly encountered in the locality, due to the deep-rootedness of traditional religion. Among the Lobi, on the other hand, descent is matrilineal.

Nevertheless, at village level, there are chiefs who perform purely administrative roles. Often referred to as "delegates", the Village

Development Committees (CVD) are auxiliaries to the deconcentrated territorial administrations. With the exception of the latter, religious authority is assured by land chiefs or "têngan-sôb" in the Dagara national language. They are responsible for all rites relating to the traditional management of land as a means of production, maintaining social order and protecting the environment. The land chief's authority may extend over several villages. His territory of influence most often corresponds to that conquered by the first occupants, generally a family or clan. This power is transmitted to descendants in a hierarchical order, from brother to brother (starting with the eldest to the youngest), then to sons, from the eldest to the youngest. It is transmitted only to men. The land chief manages his power in collaboration with other customary chiefs, including the "koumbè-sob".

In terms of religion, the peoples of the South-West are very attached to ancestor worship. Ritual sacrifices are organized periodically, according to 63.1% of the people surveyed at the study sites. These are requests for various things: good rainy seasons, health, prosperity, peace, fertility, etc., or thanks to ancestral spirits or intermediary deities. These ritual practices take place in sacred spaces (woods, trees, rivers, hills, etc.). In Dagara country, the clan is always associated with a totemic animal, usually wild, which seems to ensure its protection. This societal organization helps to protect wildlife. Nevertheless, it could also have perverse effects on wildlife biodiversity, insofar as if a clan is highly represented in a locality, its totemic animal will be extremely protected, to the detriment of all other wild animal species. Initiation ceremonies are also organized: the 'Bawr' among the Dagara and the

'Djoro' among the Lobi. The aim of these initiations is to train the whole human being (physically, morally, spiritually and intellectually).

There are also revealed religions, namely Catholicism (17.7%) as the second most represented religion in the region, followed by Islam (12.) according to the 2006 RGPH (see table 12). Catholicism and Islam are most widely practiced in urban centers, while animism is dominant in rural areas. However, the practice of these religions is often associated with that of traditional rites.

Table12 : Population distribution (in %) by province and religion

Province of residence in 2006	Animist	Muslim	Catholic	Protestant	Other	No religion	Total
Bougouriba	61,8	22,3	11	3,8	0,6	0,5	100
Poni	74,8	14	5,5	4,5	0,7	0,7	100
Ioba	50,5	6,9	40,3	1,1	0,5	0,7	100
Noumbiel	72,5	12,3	10,5	3,9	0,2	0,7	100
Set	64,9	12,9	17,7	3,3	0,6	0,7	100

Source: Monographie de la région du Sud-Ouest, 2006

3.3.3. Migratory movements

Despite its agro-pastoral potential, the South-West region has long remained "empty" for a variety of reasons. The low human occupation of the volta valleys was highlighted as a health issue, and led to the implementation of the onchocerciasis program to develop these valleys for agricultural production. According to Cambrezy & Sangli (2011), the reclamation of areas endemic to river blindness, which began in 1973, enabled populations from the central and northern parts of the country to settle thanks to the implementation of a hydro-agricultural development policy by the authorities. The aim of this policy was to settle families on developed sites in order to solve the problem of degradation of natural resources (land) in the country's most densely populated areas. But the influx to this attractive area in terms of agro-pastoral potential will cause other uncontrolled migratory movements that will accelerate the degradation of the natural resources of developed and undeveloped sites due to the diversity of agro-pastoral practices; because according to Palé (2015), spontaneous migrants oriented themselves towards villages traditionally located on the edge of developed areas because access to the lands of the villages of the Autorité de l'Aménagement des Vallées des Voltas (AVV) was strictly contingent. Migrant farming practices, characterized by the opening up of vast fields, often have a knock-on effect on indigenous populations, who used to practice mainly subsistence farming. These farming practices result in the systematic destruction of wooded areas.

Migration flows to the South-West region were to continue and intensify as a result of the major droughts of 1973-1974 and 1983-1984 in the Sahel, which had worsened agro-pastoral production conditions in the country's central and northern regions. Also, the Ivorian crisis of 2000 and 2011, which developed and worsened against a backdrop of land and ethnic conflicts (Hema & *al.,* 2018), led to a massive return of Burkinabè from Côte d'Ivoire, who settled in the western and south-western parts of the country to take advantage of climatic conditions favorable to agricultural production.

The mining boom in the early 2000s produced another type of internal mobility (Dabiré, 2016). The discovery of gold in the Bontioli wildlife reserves attracted young men and women from former gold mining sites and from various parts of the country in search of the precious metal.

Migratory movements from the study area are mainly towards the country's main urban centers (Ouagadougou and Bobo-Dioulasso) and neighboring countries (Ghana and Côte d'Ivoire). According to the Manoa village chief, the eviction of people from the partial reserve has led to temporary and permanent migration of young people to neighbouring Ghana, where they have been welcomed.

The rural exodus mainly concerns young people (men and women), whose reasons are to be found in rural development issues and technical unemployment. In these areas, farming activities last only five to six months, corresponding to the rainy season. The lack of developed lowlands, raised by almost all those interviewed, could represent a major obstacle to young people remaining on their land during the off-season. Finally, social reasons drive them to migrate in order to escape the socio-cultural burdens that weigh on them. Rural exodus is therefore a phenomenon that robs the rural environment of its lifeblood (IUCN, 2009).

The impact of the gradual increase in the population of the study sites on the plant formations is the fragmentation of the reserves through agricultural, pastoral and gold mining pressure. Indeed, these protected areas have become the preferred sites for a variety of human activities.

Villages have been set up in partial reserves, with harmful consequences for biodiversity. This observation is borne out by a study conducted by Kaboré & *al.* (2019), which reveals the existence of a positive correlation between population densities and the state of forest fragmentation. The greater the population pressure, the greater the degradation through fragmentation.

The biophysical and human characteristics of the Bontioli partial and total wildlife reserves have been analyzed in this chapter. They show an overall flat relief, with discontinuities in places marked by hills and small valleys that direct water flow towards the main rivers (the Mouhoun and Bougouriba). Soil structure is linked to the topography of the land. The vegetation is made up of a mosaic of savannah stands, characterized by degraded shrub savannah on hills and armoured plateaus, wooded savannah on soils of medium to high agronomic value, and gallery forest on the edges of watercourses.

The populations, mostly farmers, have established a non-centralized social structure based on the clan and the family. They are very attached to spirituality. Population densities have increased in recent decades due to the intensification of gold mining migration and higher birth rates.

CHAPTER 4: LAND USE AND VEGETATION CHARACTERIZATION

The study of land use was based on the interpretation of multi-date satellite images: 1986, 2002 and 2018. The aim was to determine the state of land use in the two reserves over the three time scales, and to assess and analyze the changes that have taken place in each of the reserves over the different periods. Vegetation characterization was based on the results of floristic inventories. This involved describing and analyzing the richness and diversity of the flora, as well as the structural parameters of the reserves' plant formations.

4.1. Status of reserve land use in 1986, 2002 and 2018

Land use units vary from year to year, depending on the status of the protected area.

4.1.1. Land use status of the total reserve in 1986

In 1986, the total reserve was mainly occupied by fields and fallow land, and shrub savannah. These units accounted for 60.and 31.of its total area respectively. Fields and fallow land were widespread throughout the reserve. Gallery forest covered 4.. It bordered several watercourses to the north, northeast and southeast. As for the wooded savannah, it was scattered throughout the reserve and occupied only 3.87% of the reserve's surface area. Bare soil was almost non-existent.

4.1.2. Land use status of the total reserve in 2002

Shrub savannah and wooded savannah were the most representative units in terms of total reserve area in 2002. Scattered throughout the reserve, shrub savannah covered 52.30% and wooded savannah, which was very present in the western, south-western and southern parts, occupied 39.of the reserve's total area. Fields and fallow land were located mainly to the north and west of the reserve in the wooded savannah. They covered 3.86% of the reserve. Gallery forest was present in the northern, northeastern and southeastern zones along some watercourses. It covered 4.of the reserve.

4.1.3. Land use status of the total reserve in 2018

In 2018, the total reserve was mainly characterized by shrub savannah (74.85%), wooded savannah (17.) and bare ground (5.) of the total reserve area. Gallery forest accounted for only 2.26%, and fields and fallow land for 0.. Map 1 shows the land-use units of the total reserve in 1986, 2002 and 2018.

Map1 : RTFB land-use units in 1986, 2002 and 2018

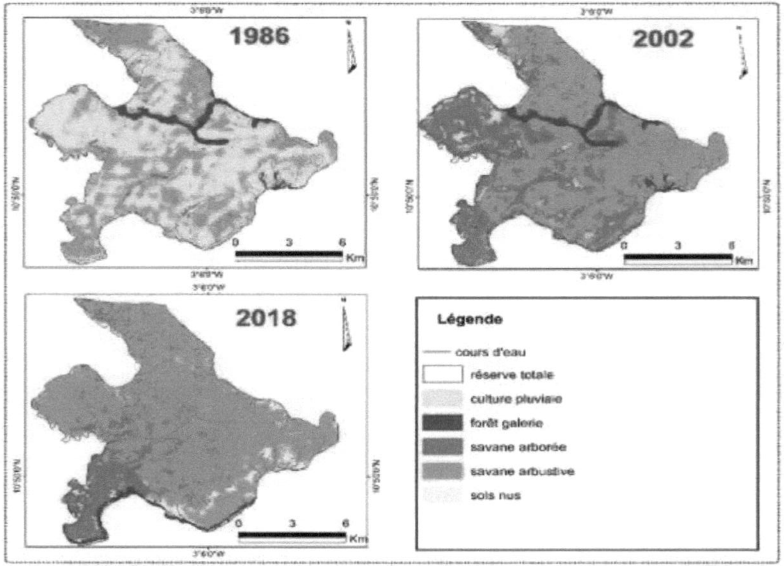

Source: IGB/BNDT 2012/Images Landsat Fevrier 2024 SOME N

4.1.4. Land use status of the partial reserve in 1986

In 1986, the partial reserve was dominated by fields and fallows (63.) and shrub savannah (29.). These two units were widespread throughout the reserve. The gallery forest along the main watercourses covered 4.90% of the reserve. Wooded savannah was rare (1.74%). It was located to the northwest and northeast. There was very little bare soil.

4.1.5. Land use status of the partial reserve in 2002

In 2002, the partial reserve was dominated by shrub savannah (52.), wooded savannah (25.93%) and fields and fallow land (16.44%). Fields and fallow land are mainly found in the contact zones between the reserve and the riverside villages. Gallery forest, which accounted for 4.92%, was mainly located in the beds of certain watercourses. Bare soil was almost non-existent.

4.1.6. Status of partial reserve land use in 2018

The partial reserve in 2018 was dominated by shrub savannah covering 53., bare soil covering 26.of the total area, and fields and fallow occupying 10.. Treed to wooded savannah covered 4.78%; the other units, gallery forest and habitat, accounted for 3.and 1.64% respectively of the reserve's total area. Map 2 shows the land-use units of the partial reserve in 1986, 2002 and 2018.

Map2 : RPFB land-use units in 1986, 2002 and 2018

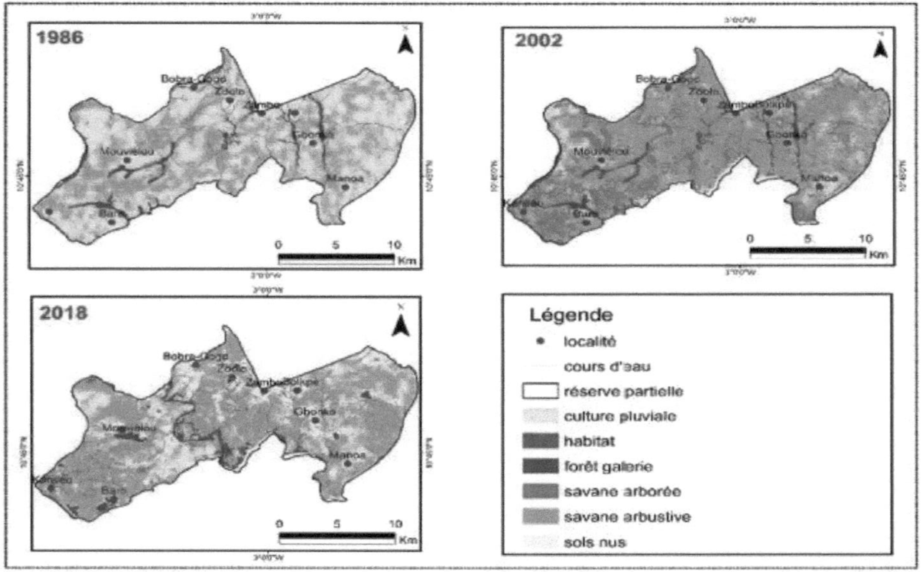

Source: IGB/BNDT 2012/Images Landsat Février 2024 SOME N.

In 1986, both reserves were invaded by fields and fallow land, despite their different statuses. In fact, in the partial reserve, usufructuary rights are granted to riparian populations. Article 6 of decree no. 3417 SE/EF establishing this reserve specified that the autochnones of Zambo and Bapla townships within the partial reserve are recognized as having customary rights to clear, cultivate, cut wood for personal use, herd grazing, fish, gather fruit and food, medicinal and dye plants subject to compliance with the regulations in force (MEECCV, 2017). These rights therefore do not open the door to the depredation of the resource insofar as the classification decree provides a framework for its exploitation.

On the other hand, the rules are very strict when it comes to the total reserve. All harvesting is forbidden, despite the fact that most of this reserve was occupied by fields. There are several reasons why fields were so common in these reserves at the time. In addition, in the 1970s, the Volta Valleys Development Program led by the State in collaboration with WHO, which helped to free areas from tsetse fly infestation, had settled agricultural migrants in the developed Bougouriba valley, which adjoins the total reserve. This situation led to the installation of clandestine agricultural migrants in the same valley, but outside the developed area, following the droughts of 1983 and 1984, which triggered intense migratory flows from the northern and central parts of the country towards these localities, which were ideal for agricultural activities. The country did not yet have a Forestry Code. Notwithstanding this, there had been a national policy for the management of classified forests since 1981, which could constitute a bulwark for the protection of these reserves, but they had never been managed.

Figure 7 shows the areas of the total reserve's land-use units in 1986, 2002 and 2018. It shows that all the land-use units of this reserve have undergone more or less significant progressive and/or regressive change over the periods 1986-2002, 2002-2018 and 1986-2018.

Figure7 : Areas of RTFB land-use units in 1986, 2002 and 2018

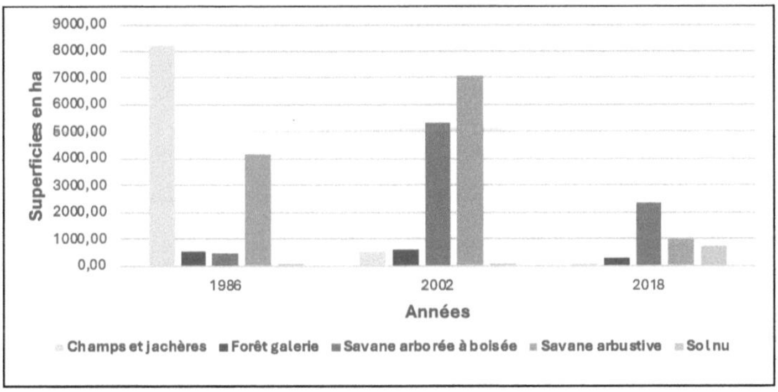

Source: Landsat images 1986, 2002 and 2018

The graph shows a predominance of fields and fallow land, wooded savannah and shrub savannah in the total reserve in 1986, 2002 and 2018 respectively.

On the other hand, in the partial reserve, the evolution is similar to that of the total reserve. Fields and fallows were the most dominant unit in 1986. By contrast, in 2002 and 2018, shrub savannah dominated the reserve (figure 8).

Figure8 : RPFB land-use unit areas in 1986, 2002 and 2018

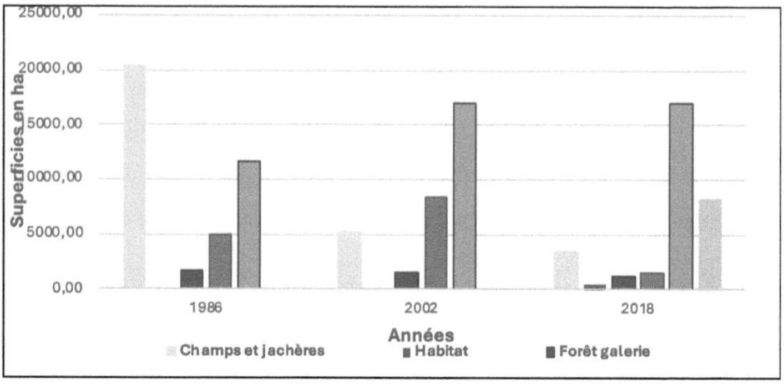

Source: Landsat images 1986, 2002 and 2018

4.2. Changes in land use on total and partial reserves

The evolution of land-use units in the two reserves is closely linked to natural and human conditions. The dynamics of land-use units, analyzed over three significant dates 1986, 2002, and 2018, has enabled us to observe significant transformations. Gains and losses of units are recorded over the different periods.

4.2.1. Changes in total reserve land use between 1986 and 2002

Fields and fallow land decreased by 57.11% between 1986 and 2002, from 60.to 3.. They were converted to wooded savannah (23.47%). On the other hand, during the same period, the area of wooded savannah increased by 35.. Shrub savannah increased by 21., to the detriment of fields and fallows (-34.) and wooded savannah (-1.). Gallery forest increased slightly by 0.at the expense of fields and fallow (-0.). Table 13 shows the percentage of the total reserve's land-use matrix from 1986 to 2002.

Table13 : RTFB land-use transition matrix from 1986 to 2002

CT	C/J	FG	SA/B	Sa	Sn	Total 1986
CP	2,63	0,17	23,47	34,70	0,002	60,97
FG	0,00	4,20	0,00	0,00	0,00	4,20
SA/B	0,25	0	2,42	1,11	0,00	3,78
Sa	0,98	0,0005	13,58	16,49	0,00	31,05
Sn	0,00	0	0,002	0,0006	0,00	0,003
Total 2002	3,86	4,37	39,47	52,30	0,002	100

CT: Thematic class, C/J: Fields and fallow land, FG: Gallery forest, SA/B: Wooded savannah, Sa: Shrubby savannah, Sn: Bare soil

Source: Landsat images from 1986 and 2002

4.2.2. Changes in land use in the partial reserve between 1986 and 2002

Between 1986 and 2002, some of the land-use units in the partial reserve declined, while others increased (see Table 14). Fields and fallow land fell from 63.60% of the reserve's total area in 1986 to 16.44 %, i.e. a decline of -47.. They have essentially been transformed into wooded savannah (15.) and shrub savannah (36.). Shrub savannah increased from 29.82% in 1986 to 52.in 2002, i.e. a total increase of 22.88% at an annual rate of 1.. This increase was achieved at the expense of wooded savannah (-0.52%), gallery forest (-0.), and fields and fallow land (-36.). Wooded savannah increased by 24., an annual rate of 1.. This growth was achieved at the expense of fields and fallow land (-15.) and gallery forest (-0.). Gallery forest changed very little over this period. It increased by only 0..

Table14 : RPFB land-use transition matrix, 1986-2002

CT	C/J	FG	SA/B	Sa	Sn	Total 1986
CP	11,93	0,07	15,50	36,10	0,003	63,60
FG	0,001	4,83	0,03	0,03	0	4,90
SA/B	0,4	0	0,8	0,52	0	1,68
Sa	4,14	0,03	9,60	16,04	0,001	29,82
Sn	0,001	0	0,0005	0,0006	0	0,002
Total 2002	16,44	4,92	25,93	52,70	0,004	100

CT: Thematic class, C/J: Fields and fallow land, FG: Gallery forest, SA/B: Wooded savannah, Sa: Shrubby savannah, Sn: Bare soil

Source: Landsat images from 1986 and 2002

4.2.3. Analysis and discussion of changes in land-use units on the two reserves between 1986 and 2002

The loss in the surface area of fields and fallow land in the two reserves can be explained by political actions in favor of the conservation of these protected areas. The implementation of the environmental policy commitments made by the Burkina Faso government on April 22, 1985, known as the "Three Struggles", i.e. the fight against abusive logging, the fight against roaming animals and the fight against bush fires, justifies the considerable reduction of fields and their conversion into shrub savannah and the conversion of shrub savannah into wooded savannah. In 1996, the State also proceeded with the eviction of the population illegally settled in these forest enclaves, and continued and strengthened its long-standing efforts to restore the vegetation cover through reforestation, monitoring and public awareness campaigns. The presence of large areas of wooded savannah in the total reserve is proof that there is less anthropogenic activity within it (Somé, 2013). In addition to these protective measures, it is worth noting that, according to several authors (Ouoba, 2013; Nouaceur, 2020), rainfall has picked up again after the droughts of 1983, 1984 and 1985 in the Sahel and, by extension, in Burkina Faso, especially since the 1990s. This observation is also borne out by Koné & *al* (2007), who show that the 1990-1999 decade was marked by a slight upturn in rainfall, accompanied by a recovery in vegetation cover in the Sudanian savannah of Côte d'Ivoire; this would have had a favorable impact on vegetation development. The combination of these two parameters led to the abandonment of fields in the reserves and the return of some migrants to their areas of departure, and to a resumption of vegetation growth, encouraging the mutation of certain plant formations.

4.2.4. Changes in total reserve land use between 2002 and 2018

Shrub savannah has increased in area (22.54%), rising from 52.32% in 2002 to 74.in 2018. The annual rate of increase is 1.. This increase came at the expense of wooded savannah (-26.28%), gallery forest (-3.79%) and fields and fallow (-2.98%). The area of bare soil has grown spectacularly; from almost non-existent in 2002 (0.002%), it has risen to 5.28% in 2018, an evolution of 5.corresponding to an annual rate of 32.94%. Bare soil increased at the expense of other units, namely shrub savannah (-3.33%), wooded savannah (-1.34%), gallery forest (-0.) and fields and fallow land (-0.).

On the other hand, other units regressed over the same period. These include wooded savannah, whose area fell from 39.46% in 2002 to 17.in 2018, representing a regression rate of -22.21%. This unit has regressed annually by -1.to the benefit of shrub savannah (6.44%). Gallery forest lost 2.of its surface area to wooded savannah (1.) and shrub savannah (0.). Fields and fallow land declined by -3.(-0.p.a.). Shrub savannah (0.) and wooded savannah (0.). Table 15 shows the percentage of the total reserve's land-use matrix from 2002 to 2018.

Table15 : RTFB land-use transition matrix from 2002 to 2018

CT	C/J	FG	SA/B	Sa	Sn	Total 2002
CP	0,03	0,05	0,52	2,98	0,27	3,86
FG	0,005	0,00	0,24	3,79	0,34	4,37
SA/B	0,15	1,63	10,06	26,28	1,34	39,46
Sa	0,17	0,58	6,44	41,79	3,33	52,31
Sn	0,00	0,00	0,00	0,001	0,0006	0,002
Total 2018	0,36	2,26	17,25	74,85	5,28	100

CT: Thematic class, C/J: Fields and fallow land, FG: Gallery forest, SA/B: Wooded savannah, Sa: Shrubby savannah, Sn: Bare soil
Source: Landsat images from 2002 and 2018

4.2.5. Changes in land use in the partial reserve between 2002 and 2018

The land-use transition matrix for the partial reserve from 2002 to 2018 reveals significant unit mutations (see Table 16). Treed to wooded savannah declined by -21.. By contrast, bare soil, which was virtually non-existent (0.002%) in 2002, has increased by . Shrub savannah (16.), wooded savannah (4.), fields and fallow land (4.) and gallery forest (1.) were converted to bare soil. They decreased by -5.. Habitat occupied 1.in 2018. Gallery forest and shrub savannah declined slightly by -1.19% and -0.34% respectively.

Table16 : RPFB land-use transition matrix from 2002 to 2018

CT	C/J	FG	Hab	SA/ B	Sa	Sn	Total 2002
CP	3,90	0,28	0,38	0,40	7,08	4,39	16,45
FG	0,24	0,10	0,03	0,27	2,96	1,31	4,92
SA/B	2,61	1,96	0,44	2,45	14,25	4,22	25,93
Sa	4,04	1,39	0,79	1,66	28,74	16,09	52,70
Sn	0,00 1	0,00 07	0,00	0,00 02	0,001	0,0008	0,002
Total 2018	10,8 0	3,73	1,64	4,78	53,04	26,01	100

CT: Thematic class, C/J: Fields and fallow land, FG: Gallery forest, SA/B: Treed to wooded savannah, Sa: Shrubby savannah, Sn: Bare soil, Hab: Habitat

Source: Landsat images from 2002 and 2018

4.2.6. Analysis and discussion of changes in land-use units on the two reserves between 2002 and 2018

The decline in wooded savannah and gallery forest is due to clandestine anthropogenic activities such as greenwood cutting, gold panning and livestock farming. The consequence of the anthropization of gallery forest is the erosion of river banks. The increase in the surface area of bare soil can be attributed to these activities. In fact, gold panning has developed in these reserves over the last two decades, so its impact on plant formations may only be very noticeable in recent years. In the field, the scene is one of desolation (scattered holes). Pastoral practices are also destructive to plant formations. Large trees, whose leaves are prized by livestock, are pruned. This is the case of *Afzelia africana, Pterocarpus erinaceus and Khaya senegalensis.* Shrubs are encroached upon by animals, especially around waterholes, creating haloes of bare ground in these areas. Since wood is the main fuel used by local households (99.1%), the reserves represent a reservoir of fuelwood for 56.9% of those questioned. Abusive logging contributes to the increase in bare areas. The corollary is that bare soil is exposed to erosion and runoff. These phenomena strip the topsoil from the soil, while impoverishing it (Yaméogo & *al.*, 2009). Finally, the decline in rain-fed cultivation is explained by the protection measures undertaken by the State and its decentralized services.

In the partial reserve, the exacerbated progression of bare soil is justified for several reasons:

- the slash-and-burn agricultural production system exposes the soil to rapid impoverishment followed by accelerated degradation. The same is true of the farming practices of some migrants, which consist in completely clearing shrubs and trees, including food trees.

- the spectacular development of harmful pastoral practices. Herders settle in the reserves all year round. The immediate consequence of these bare zones could be the appearance acceleration of water and wind erosion phenomena. Bare soil can also characterize burnt grassy savannahs and rocky outcrops (Koné & *al.,* 2007).

the spectacular development of gold panning in these reserves.

4.2.7. Changes in total reserve land use between 1986 and 2018

From 1986 to 2018, the land-use units of the total reserve have undergone profound changes. Area losses and gains were recorded at all levels. Bare soil gained in area (5.27%). Fields and fallow land, shrub savannah and gallery forest were converted into this unit for 3.,

1.and 0.respectively. Shrub savannah is the main unit in terms of surface cover in 2018 with 74.. Its surface area has increased by 43.80% in 32 years, representing average annual growth of 1.. The expansion of this plant formation has been at the expense of fields and fallow land (-46.), gallery forest (-3.) and wooded savannah (-2.). The area of wooded savannah increased by 13., for an annual rate of 0.. Area losses were recorded in fields and fallow land (-60.61%), which fell from 60.97% in 1986 to 0.in 2018; an annual regression of -1.. The regression was to the benefit of wooded savannah (8.68%), shrub savannah (46.), bare soil (3.) and gallery forest (1.). Finally, gallery forest lost 1.94% of its area to shrub savannah (0.) and wooded savannah (0.). Table 17 shows the percentage of the total reserve's land-use transition matrix from 1986 to 2018.

Table17 : RTFB land-use transition matrix from 1986 to 2018

CT	C/J	FG	SA/B	Sa	Sn	Total 1986
CP	0,25	1,56	8,68	46,74	3,74	60,97
FG	0,005	0,00	0,24	3,65	0,31	4,20
SA/B	0,003	0,06	0,85	2,80	0,05	3,77
Sa	0,11	0,64	7,47	21,65	1,17	31,04
Sn	0,00	0,00	0,002	0,0006	0,00	0,002
Total 2018	0,36	2,26	17,24	74,84	5,28	100

CT: Thematic class, C/J: Fields and fallow land, FG: Gallery forest, SA/B: Wooded savannah, Sa: Shrubby savannah, Sn: Bare soil
Source: Landsat images from 1986 and 2018

4.2.8. Changes in land use in the partial reserve between 1986 and 2018

The land-use transition matrix for the partial reserve from 1986 to 2018 shows remarkable transformations in the various units. Fields and fallow land, which occupied 63. of the reserve's area in 1986, fell to 10.in 2018. They decreased by 52.for an annual proportion of -1.. The area lost to fields and fallow has been converted into 31.96% shrub savannah, 18.12% bare ground, 2.wooded savannah and 2.gallery forest. Shrub savannah (23.22%) has recovered at a rate of 0.per year, to the detriment of gallery forest (-3%) and wooded savannah (-1.11%). Wooded savannah increased by 3.09%, while 2.of this plant formation was converted to shrub savannah. Gallery forest decreased overall by -1.. On the other hand, it gained 1.and 0.at the expense of shrub savannah (-1.45%) and wooded to wooded savannah (-0.). The extension of bare soil was at the expense of fields and fallow land (-18.), shrub savannah (-6.) and gallery forest (-1.27%). Table 18 presents the land-use transition matrix of the partial reserve in percentages from 1986 and 2018.

Table18 : RPFB land-use transition matrix from 1986 to 2018

CT	C/J	FG	Hab	SA/B	Sa	Sn	Total 1986
CP	7,84	2,05	1,30	2,33	31,96	18,12	63,60
FG	0,23	0,10	0,03	0,27	3,00	1,27	4,90
SA/B	0,10	0,13	0,006	0,14	1,11	0,20	1,69

Sa	2,63	1,45	0,31	2,04	16,96	6,42	29,81
Sn	0,00	0,0001	0,00	0,00	0,001	0,0005	0,002
Total 2018	10,80	3,73	1,64	4,78	53,03	26,01	100

CT: Thematic class, C/J: Fields and fallow land, FG: Gallery forest, Hab: Habitat, SA/B: Wooded to wooded savannah, Sa: Shrubby savannah, Sn: Bare soil

Source: Landsat images from 1986 and 2018

4.2.9. Analysis and discussion of the evolution of land-use units in the two reserves between 1986 and 2018

In both reserves, bare soil is the unit that has progressed the most, while fields and fallow land are those that have regressed the most.

Bare soil has appeared mainly in areas vulnerable to degradation, notably in cultivated areas and shrub savannah. Low vegetation cover, combined with various human pressures, explains the transformation of wooded areas into bare soil. In arid or semi-arid zones, the spaces left by tree felling can be subject to extreme thermal conditions, favoring the development of xeric species or the extension of bare, crusted spaces (Sawadogo, 2009). These bare soils are exposed to the impact of water and wind erosion (Ouoba & *al.,* 2014). Furthermore, the growth in the area of shrub savannah is justified by the fact that an agricultural area abandoned over the course of a few decades, but subject to other anthropogenic pressures, is transformed into a bushy formation.

Ultimately, the evolution of plant formations in the two reserves is linked to the incoordination of actions to protect and exploit the resources of these protected areas. Moreover, when state protection measures are stepped up through projects and programs, the result is that plant formations evolve towards more wooded units, and vice versa when these measures come to an end.

4.3. Characteristics of reserve plant formations

The characteristics of the reserves' plant formations are analyzed through floristic composition, tree vitality, floristic diversity and structural parameters.

4.3.1. Floristic richness

The term floristic richness refers to several notions, namely specific richness, generic richness and relative abundance.

4.3.1.1. Floristic richness of the total reserve

4.3.1.1.1. Specific richness

The flora of the Bontioli total fauna reserve, according to the results of the floristic inventory, is composed of 49 species, grouped into 36 genera and 18 families, the most representative of which, in numerical order, are Combretaceae (25,44%), Fabaceae (14.69%), Sapotaceae (12.), Rubiaceae (11.54%), Ebenaceae (11.), Cesalpiniaceae (7.89%), Anacardiaceae (6.), Mimosaceae (4.), Annonaceae (2.72%) and Meliaceae (2.). Less represented or rare families include Bombacaceae, Balanitaceae, Loganiaceae, Ulmaceae, Apocynaceae, Olacaceae, Sterculiaceae and Verbenaceae. Figure 9 shows the systematic spectrum of this reserve.

Figure9 : Systematic spectrum of RTFB

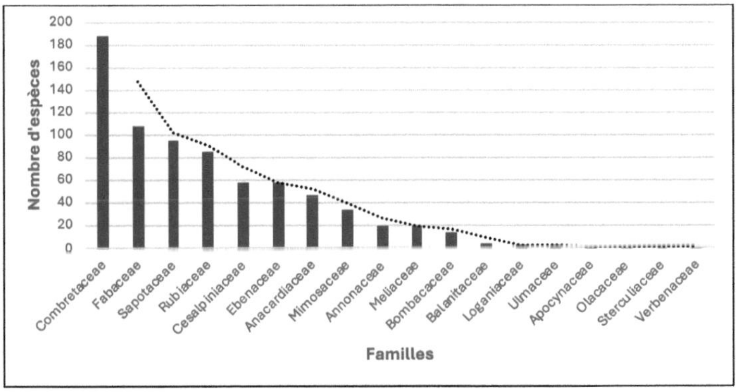

Source: Floristic inventory data, December 2020

4.3.1.1.2. Factors in the distribution of families

The distribution of families by plant formation unit shows that gallery forest is the richest, with 16 families, compared with 8 families for agroforestry/ fallow park, which is the least rich plant formation. For the other formations, namely wooded savannah, tree savannah and shrub savannah, there were 12, 10 and 13 families respectively. The poverty of the agroforestry/fallow park is justified by the anthropic pressure it has been subjected to. In this cropping system, clearing and cultivation means sparing mainly "utilitarian" plant species, which leads to the deliberate destruction of certain plant species and, in turn, their families. Table 19 shows the distribution of plots, number of individuals, species, genera and families according to plant formations in the total reserve.

Table19 : Distribution of plots, number of individuals, species, genera and families according to vegetation units in the RTFB.

Plant formation	Number of plots	Number of individuals	Number of species	Number of types	Number of families
Gallery forest	4	188	27	26	16
Wooded savannah	3	135	24	21	12
Wooded savannah	6	210	27	20	10
Shrubby savannah	5	124	24	21	13
Agroforestry park / Fallow land	3	78	13	11	8
Total reserve	21	735	49	36	18

Source: Floristic inventory data, December 2020

The gallery forest is dominated by Combretaceae (35.), Ebenaceae (27.) and Fabaceae (16.). In the wooded savannah, the most important families are Fabaceae (22.34%), Sapotaceae (21.) and Cesalpiniaceae (7.). The tree savannah is mainly composed of Rubiaceae (24.76%), Combretaceae (21.43%), Cesalpiniaceae (15.24%), Fabaceae (9.05%), Mimosaceae and Anacardiaceae (7.). In the shrub savannah, the most common families are Combretaceae (30.65%), Anacardiaceae (20.), Rubiaceae (14.52%) and Annonaceae (11.). Finally, in the agroforestry/fallow park, Combretaceae and Sapotaceae (35.), Fabaceae (15.) and Mimosaceae (5.) are the most representative.

Statistics show that no single family is absolutely dominant in all plant formations in this reserve. Nevertheless, Combretaceae remains the most represented family in almost all plant formations.

4.3.1.1.3. Generic richness

Specific quotients are low for all plant formations in the total reserve: 1.14 for wooded and shrub savannahs, 1.04, 1.35 and 1.18 for gallery forest, wooded savannah and agroforestry/ fallow park. The overall specific quotient (SQ) for the reserve is 1.36. This indicates that the reserve's various plant formations are evolving towards stability.

4.3.1.1.4. Relative abundance

The most abundant plant species in the total reserve are, in order of importance, *Anogeissus leiocarpus* (17.), *Vitellaria paradoxa* (12.), *Crossopteryx febrifuga* (10.), *Diospyros mespiliformis* and *Pterocarpus erinaceus* (7.). In the various plant formations of the reserve, they are distributed according to Table 20, which summarizes some of the most abundant species. The most abundant species by type of plant formation are *Anogeissus leiocarpus* (44.68%), *Crossopteryx febrifuga* (23.33%), *Vitellaria paradoxa* (29.23%), *Lannea microcarpa* (17.74%) *and Vitellaria paradoxa* (35.) for gallery forest, wooded savannah, wooded savannah, shrub savannah and agroforestry park/ fallow land respectively.

Table20 : Distribution of the most abundant species according to plant formations in the RTFB

Plant formations	Species	Relative abundance (%)
Gallery forest	*Anogeissus leiocarpus*	44,68
	Diospyros mespiliformis	27,13
	Pterocarpus erinaceus	5,32

Wooded savannah	*Crossopteryx febrifuga*	23,33
	Detarium microcarpum	11,90
	Terminalia avicennioides	8,57
	Vitellaria paradoxa	8,57
	Combretum fragrans	6,67
	Lannea microcarpa	5,71
Wooded savannah	*Vitellaria paradoxa*	29,63
	Pterocarpus erinaceus	17,78
	Anogeissus leiocarpus	13,33
Shrubby savannah	*Lannea microcarpa*	17,74
	Combretum glutinosum	12,10
	Crossopteryx febrifuga	12,10
	Hexalobus monopetalus	11,29
	Terminalia macroptera	8,87
Agroforestry park/ fallow land	*Vitellaria paradoxa*	35,90
	Anogeissus leiocarpus	24,36
	Pterocarpus erinaceus	15,38
	Combretum fragrans	7,69

Source: Floristic inventory data, December 2020

4.3.1.2. Floristic richness of the partial reserve

4.3.1.2.1. Specific richness

The flora of the partial reserve is relatively more diverse than that of the total reserve. It comprises 59 species, grouped into 42 genera and 20 families according to the results of the floristic inventory. The most important families in numerical terms are Combretaceae (21.01%), Anacardiaceae (15.21%), Mimosaceae (12.), Sapotaceae (11.), Rubiaceae (8.74%), Ebenaceae (6.89%), Cesalpiniaceae (5.), Annonaceae (5.29%), Meliaceae (3.78%), Balanitaceae (2.29%) and Fabaceae (2.). Figure 10 shows the systematic spectrum of the partial rcscrvc.

Figure10 : Systematic spectrum of RPFB

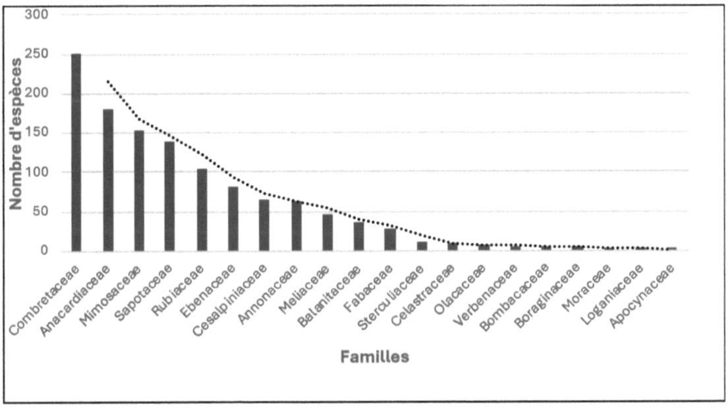

Source: Floristic inventory data, December 2020

Analysis of the floristic composition of the different plant formation units shown in Table 21 reveals that the richest unit is the shrub savannah (16 families) and the poorest is the gallery forest (12 families). Degradation of the Bougouriba riverbanks by pastoral activities (herd watering, tree pruning), gardening, fuelwood and timber cutting are among the factors responsible for this situation. The shrub savannah and the agroforestry/ fallow park are the most represented units in terms of land use in the reserve, and have the most diverse flora, with 36 species, 16 and 14 families respectively. The table below shows the distribution of plots, number of individuals, species, genera and families according to plant formations in the reserve.

Table21 : Distribution of plots, number of individuals, species, genera and families according to vegetation units in the RPFB.

Plant formations	Numb er of plots	Number of individu als	Numb er of specie s	Num ber of types	Numbe r of families
Gallery forest	4	171	22	18	12
Wooded savannah	3	193	28	21	13
Wooded savannah	6	295	30	20	13
Shrubby savannah	10	308	36	25	16
Agroforestry park/fallowland	10	223	36	26	14
Partial reserve	33	1190	59	42	20

Source: Floristic inventory data, December 2020

In the gallery forest, the most important families are Ebenaceae (27.49%), Combretaceae (20.), Balanitaceae (12.), Mimosaceae (10.), Rubiaceae (9.94%) and Meliaceae (5.). In the wooded savannah, Rubiaceae (22.), Anacardiaceae (22.), Combretaceae (17.) and Sapotaceae (13.) are the most important families. In the wooded savannah, Combretaceae (29.80%), Anacardiaceae (21.), Annonaceae (16.) and Mimosaceae (12.) are the most important families. In terms of specific importance, the shrub savannah is made up respectively of

Anacardiaceae (16.88%), Combretaceae and Mimosaceae (15.91%), Sapotaceae (12.34%), Cesalpiniaceae (9.42%), Meliaceae (8.12%), Rubiaceae (6.82%) and Ebenaceae (5.). Finally, the agroforestry/fallow park is mainly composed of Sapotaceae (29.15%), Combretaceae (19.), Mimosaceae (18.), Anacardiaceae (7.) and Rubiaceae (7.).

4.3.1.2.2. Factors in the distribution of families

The plant formations in the partial reserve are made up of several families. Some of these families are common to all plant formations. There are several reasons for this: the almost homogeneous climatic and edaphic conditions that are responsible for their establishment and development. As the surveys were carried out on transects a few kilometers long, edaphic conditions are somewhat similar, even if precautions have been taken to minimize this. The modification of the forest area by the population, with the appearance of units such as agroforestry parks and certain shrub savannahs, which are a reconstitution of vegetation after human exploitation.

Combretaceae is the most important family and is represented in all plant formations in the partial reserve. Species from this family are less subject to pastoral pressure, as they are generally less palatable (Bognounou, 2004). Another major fact is the dominance of Sapotaceae and Mimosaceae in the agroforestry/fallow park. In this plant formation, these families are dominated respectively by *Vitellaria paradoxa* (Sapotaceae), *Parkia biglobosa* and *Faidherbia Albida* (Mimosaceae). These species are generally spared when land is cleared because of their usefulness for food, economic, pastoral and agricultural purposes. Farmers prefer species like *Faidherbia albida* because they fix nitrogen from the air and help fertilize the soil, and they are also highly valued by animals (Savadogo & *al.,* 2015). These same observations of farmers' recognition of certain species, in this case *Faidherbia albida*, in improving soil fertility were made in the Mayahi department of Niger (Boubacar, 2010) and in the Centre-West of Burkina Faso in the rural commune of Imasgo by Kaboré (2018). The desire of populations to conserve *Parkia biglobosa*, *Vitellaria paradoxa* and *Tamarindus indica* in their farming systems was clearly highlighted in northern Benin by Sounon Bouko & *al.* (2007) and in the region of Tandjilé-Ouest in Chad by Avana-Tientcheu & *al.* (2019).

4.3.1.2.3. Generic richness

As far as specific quotients are concerned, they are low in all plant formations in both the partial and total reserves. In order of importance, they are 1.22, 1.33, 1.38, 1.44 and 1.50 respectively for gallery forest, wooded savannah, agroforestry park/ fallow, shrub savannah and tree savannah. The overall specific quotient of the partial reserve is 1.40.

The plant formations on both reserves have specific quotients with values close to 1, characteristic of old-growth stands.

4.3.1.2.4. Relative abundance

The partial reserve contains a diversity of plant species, but the most abundant are *Lannea microcarpa* (13.28%), *Vitellaria paradoxa* (11.34%), *Anogeissus leiocarpus* (8.), *Crossopteryx febrifuga* (6.64%), *Diospyros mespiliformis* (6.) and *Hexalobus monopetalus* (5.).

In terms of plant formations, the most abundant taxa in the gallery forest are *Diospyros mespiliformis* (27.49%), *Anogeissus leiocarpus* (19.), *Balanites aegyptiaca* (12.) and *Mitragyna inermis* (8.); those of the wooded savannah are *Crossopteryx febrifuga* (22.80%), *Lannea microcarpa* (20.73%), *Vitellaria paradoxa* (9.84%) and *Anogeissus leiocarpus* (7.). The tree savannah includes *Hexalobus monopetalus* (16.), *Lannea microcarpa* (16.), *Anogeissus leiocarpus* (13.) and *Entada africana* (11.). In the shrub savannah, the most abundant species are *Lannea microcarpa* (15.26%), *Vitellaria paradoxa* (13.31%), *Pseudocedrela kotschyi* (7.14%), *Acacia senegal* (6.49%), *Crossopteryx febrifuga* (6.17%), *Piliostigma thonningii* (6.), and *Diospyros mespiliformis* (5.). Finally, *Vitellaria paradoxa* (29.15%), *Lannea microcarpa* (7.17%), and *Acacia macrostachya* (6.) are the most abundant species in the agroforestry/fallow park. Table 22 shows the distribution of the most abundant species according to plant formations in the partial reserve.

Table22 : Distribution of the most abundant species according to RPFB plant formations

Plant formations	Plant species	Relative abundance (%)
Gallery forest	*Diospyros mespiliformis*	27,49
	Anogeissus leiocarpus	19,30
	Balanites aegyptiaca	12,28
	Mitragyna inermis	8,19
Wooded savannah	*Crossopteryx febrifuga*	22,80
	Lannea microcarpa	20,73
	Vitellaria paradoxa	9,84

	Anogeissus leiocarpus	7,25
Wooded savannah	*Hexalobus monopetalus*	16,95
	Lannea microcarpa	16,95
	Anogeissus leiocarpus	13,22
	Entada africana	11,19
Shrubby savannah	*Lannea microcarpa*	15,26
	Vitellaria paradoxa	13,31
	Pseudocedrela kotschyi	7,14
	Acacia senegal	6,49
	Crossopteryx febrifuga	6,17
	Piliostigma thonningii	6,17
	Diospyros mespiliformis	5,52
Agroforestry park and fallow land	*Vitellaria paradoxa*	29,15
	Lannea microcarpa	7,17
	Acacia macrostachya	6,69

Source: Floristic inventory data, December 2020

Fruit species for human consumption are generally spared from clearing for agriculture. This is the case of *Vitellaria paradoxa*, a plant species protected by farmers. It is abundant in the agroforestry/ fallow park, for its economic usefulness insofar as its fruits are consumed and provide income for rural households.

Analysis of the systematic spectra of the two reserves shows that the partial reserve has a more diverse flora than the total reserve, both in terms of specific and generic components. However, the Combrétacées family remains dominant in both reserves. Apart from Combrétacées, the partial reserve is characterized by Anacardiaceae, Mimosaceae, Sapotaceae and Rubiaceae. On the other hand, in the total reserve, Fabaceae, Sapotaceae and Rubiaceae characterize it.

4.3.2. Vitality of woody plants

The vitality or state of health of a plant depends on a range of factors that are complex to identify and measure. Assessing a tree's vitality should obviously involve all three major parts: the root system, the trunk and the crown. It is an ecologically ambiguous concept in that health is not a property of the individual tree, stand or forest (Landmann, 1988). In this study, we identified five criteria for the poor health of trees: lopped, burnt, holed, semi-dried and standing dead. Any woody tree that does not meet any of these criteria is considered healthy.

The plant formations in the total reserve, namely gallery forest, wooded savannah, tree savannah, shrub savannah and agroforestry park/ fallow land, contain 82.45%, 95.56%, 88.5%, 77.4% and 70.healthy woody species respectively. In the same order, delimbed woody species are 2.48%, 0%, 2., 4.and 15.. Burnt woody species are found in tree savannah (0.48%) and shrub savannah (5.). In order of frequency, woody plants with holes were found in gallery forest (4.26%), wooded savannah (2.), shrub savannah (1.), wooded savannah (1.) and agroforestry park (1.28%). On the other hand, semi-dried ligneous plants are found in all plant formations, in varying proportions. Shrub savannah, gallery forest and wooded savannah have the highest rates, at 11.29%, 10.and 5.respectively. The wooded savannah and agroforestry/fallow park contain 2.22% and 3.of semi-dried individuals. Standing dead woody plants were found in the agroforestry/fallow park (8.) and gallery forest (0.). Table 23 summarizes these data.

Table23 : Vitality of woody plants in RTFB plant formations

Vitality	Plant formations									
	Gallery forest		Wooded savannah		Wooded savannah		Shrubby savannah		Agroforestry park/ fallow land	
	Nb/ind	%	Nb/ind	%	Nb/ind	%	Nb/ind	%	Nb/ind	%
1*	155	82,45	129	95,56	186	88,57	96	77,42	55	70,51
2	5	2,66	0	0	5	2,38	5	4,03	12	15,38
3	0	0	0	0	1	0,48	7	5,65	0	0
4	8	4,26	2	1,48	6	2,86	2	1,61	1	1,28
5	20	10,64	3	2,22	12	5,71	14	11,29	3	3,85
6	0	0	1	0,74	0	0	0	0	7	8,97
Total	188	100	135	100	210	100	124	100	78	100

*1= Sound, 2= Shaken, 3= Burnt, 4= Bored, 5= Semi-dried, 6= Dead on the ground

Source: Floristic inventory data, December 2020

Analysis of the table reveals a high proportion of lopped and standing dead trees in the agroforestry park (15.38% and 8.respectively), and virtually no standing dead trees in the other plant formations. The agroforestry/ fallow park also had the lowest proportion of healthy trees (70.51%). These figures show the damaging impact of human activity on vegetation. In addition, unhealthy techniques are used to kill standing trees. These include debarking the entire circumference of the trunk, from the base to the middle, thus preventing the sap from flowing from the roots to the branches and then to the leaves. Another technique consists, as part of field preparation, in piling crop residues (stems, leaves, etc.) and cut shrubs around the trunk and roots of the troublesome tree. This waste is then incinerated.

In the agricultural practices of the study sites, many trees are mutilated to clear fields so that seedlings from seedlings have access to light for their growth. Residents also prune trees in their fields for utilitarian purposes, such as making daba handles, ladders and firewood. There is also a high proportion of semi-dried trees in the shrub savannah (11.29%), gallery forest (10.) and tree savannah (5.). This phenomenon is linked to parasitic attacks and climatic variability (higher temperatures and lower rainfall), which could lead to a proliferation of insects harmful to plants. The leathery or shallow structure of the soil on which the shrub savannah is established, combined with the climatic factors listed above, could also be one of the reasons.

Within the partial reserve, healthy trees are distributed in numerical order of importance, i.e. wooded savannah (87.45%), shrub savannah (86.), gallery forest (87.), wooded savannah (82.) and agroforestry park (77.). Delimbed trees are found in agroforestry park/ fallow land (15.), gallery forest (9.38%), wooded savannah (8.29%), wooded savannah (5.) and shrub savannah (4.). Burnt woody species are not very representative, with rates ranging from 0.58% in gallery forest to 1.in shrub savannah. In terms of woody trees with holes, 5., 3., 2., 1.and 1.were recorded in wooded savannah, shrub savannah, wooded savannah, agroforestry park/ fallow land and gallery forest. Semi-dried ligneous trees were classified in descending order according to the plant formation concerned: wooded savannah (4.41%), agroforestry park/ fallow land (4.04%), shrub savannah (3.), wooded savannah (2.07%) and gallery forest (1.). Finally, standing dead trees are very rare in all plant formations in this reserve: gallery forest (0.) and agroforestry park/ fallow (0.45%) (see Table 24).

Table24 : Vitality of woody plants in RPFB plant formations

Vitality	Plant formations									
	Gallery forest		Wooded savannah		Wooded savannah		Shrubby savannah		Agroforestry park/ fallow land	
	Nb/ind	%	Nb/ind.	%	Nb/ind.	%	Nb/ind.	%	Nb/ind.	%
1*	149	87,13	160	82,90	258	87,45	269	87,34	173	77,58
2	16	9,38	16	8,29	16	5,42	13	4,22	34	15,25
3	1	0,58	2	1,04	2	0,68	6	1,95	2	0,90
4	3	1,75	10	5,18	6	2,03	10	3,25	4	1,79
5	2	1,17	4	2,07	13	4,41	10	3,25	9	4,04
6	1	0,58	0	0	0	0	0	0	1	0,45
Total	171	100	193	100	295	100	308	100	223	100

*1= Sound, 2= Shaken, 3= Burnt, 4= Bored, 5= Semi-dried, 6= Dead on the ground

Source: Floristic inventory data, December 2020

The table shows higher proportions of healthy individuals in all plant formations in the partial reserve than in the total reserve. This observation may be justified by the fact that this reserve is more accessible to people who carry out wood-cutting activities. Dead, semi-dried and burnt woody plants are cut as a matter of priority. The Zambo dolotières said during the focus groups: "(...) we first look for dead wood on the river banks and in the forest (...)". In terms of lopped trees, the proportions are higher in gallery forest (9.38%), wooded savannah (8.) and shrub savannah (5.) than in the same plant formations in the total reserve. This means that greenwood harvesting is a reality in all the reserve's plant formations. Since wood cutting by local people is clandestine, large trees are pruned. In addition, shepherds prune trees to boost livestock feed.

Considering the vitality criteria listed above, and according to the results of the samples taken from the two reserves, the total and partial reserves had 84.49% and 84.79% healthy woody species respectively. On the other hand, the total reserve recorded more delimbed wood (7.98%) than the partial reserve (3.). Burnt woody species recorded the same rate in both reserves (1.09%). Woody plants with holes, semi-dried and dead on the ground represent 2.59%, 7.and 1.of the total number of woody plants inventoried in the total reserve, and 2.77%, 3.and 0.of those inventoried in the partial reserve. Table 25 shows the vitality of woody plants in the two reserves.

Table25 : Vitality of woody plants on both reserves

Woody plant health	Total reserve		Partial reserve	
	Number of species	Frequency (%)	Number of species	Frequency (%)
Healthy	621	84,49	1009	84,79
Tangled	27	3,67	95	7,98
Burnt	8	1,09	13	1,09
Hole	19	2,59	33	2,77
Semi-dried	52	7,07	38	3,19
Dead on your feet	8	1,09	2	0,17
Total	735	100	1190	100

Source: Floristic inventory data, December 2020

Considering the statistics shown in the table, the vitality of the trees in both reserves is relatively good. Nevertheless, 14.and 15.of trees in the total and partial reserves respectively have low vitality, and therefore reduced growth. The existence of dislodged ligneous trees in the total reserve and their relatively high rate in the partial reserve is a poor indicator of human action on the plant formations of these protected areas. The presence of burnt woody debris indicates the existence of late fires in these forest enclaves. Abiotic factors (high temperatures and water stress), age and attacks by biotic agents (parasites) are thought to be responsible for some of the trees dying, semi-drying and with holes in them. Tree dieback is also a natural process in forest dynamics (Etzold & *al.,* 2016).

4.3.3. Specific diversity

Diversity takes into account the number of species, and the distribution of individuals within these species. It has been evaluated using various indices.

Shannon, Simpson and regularity (equitability) indices are convenient for comparing and characterizing plant formations (Somé & *al.,* 2011). For Frontier & Pichod-Viale (1991), diversity is low when $H'<3$ bits, medium if H' is between 3 and 4 bits, then high when $H'\geq4$ bits. In the total reserve, gallery forest and the agroforestry/fallow park are the least diverse plant formations, with low Shannon indices of 1.78 bits and 1.84 bits respectively. Savannahs have average diversity compared with other plant formations, but shrub savannah is the most diverse (2.68 bits), followed by tree savannah (2.63 bits), then wooded savannah (2.20 bits). On the Frontier and Pichod-Viale scales, all plant formations in the total reserve have low diversity. Indeed, maximum diversity ranges from 2.56 bits (agroforestry park) to 3.30 bits (gallery forest and wooded savannah). As for Piélou's equitability index (E), which measures the weight of each species in the stand, individuals of each species in the shrub and tree savannahs are almost evenly distributed, with index values of 0.84 and 0.80 respectively; those in the wooded savannahs and agroforestry park/fallow forest are moderately distributed (0.69 and 0.72). On the other hand, individuals of the various gallery forest species are poorly distributed, with an equitability index of 0.54. Simpson's index ranges from 0.09 to 0.28, indicating a relatively low dominance of any one species in all plant formations on this reserve. Nevertheless, this low dominance is even more pronounced in the savannah formations, where the index is between 0.09 and 0.15, than in the agroforestry park and gallery forest, where it is between 0.22 and 0.28. Finally, Simpson's inverse index is congruent with the other diversity indices (see Table 26).

Table26 : Distribution of diversity and equitability indices for RTFB plant formations

Plant formations	Shannon diversity index (H')	Maximum diversity (H'max)	Piélou equitability index (E)	Simpson Index (D)	Simpson's inverse index (1-D)
Gallery forest	1,78	3,30	0,54	0,28	0,72
Wooded savannah	2,20	3,18	0,69	0,15	0,85
Wooded savannah	2,63	3,30	0,80	0,10	0,90
Shrubby savannah	2,68	3,18	0,84	0,09	0,91
Agroforestry park/ Fallow land	1,84	2,56	0,72	0,22	0,78

Source: Floristic inventory data, December 2020

The low diversity indices observed in the gallery forest and agroforestry/ fallow park indicate the importance of a few species in these plant formations. *Anogeissus leiocarpus* and *Diospyros mespiliformis* account for 44.and 27.respectively of the total number of individuals recorded in these plant formations. In the agroforestry/ fallow park, *Vitellaria paradoxa, Anogeissus leiocarpus* and *Pterocarpus erinaceus* are the species whose individuals dominate at 35.90%, 24.and 15.respectively. These dominant species play a major role in the functioning of the ecosystem by controlling the flow of energy, but it is the rare species that condition the diversity of the stand (Rouamba, 1997). The absence of undergrowth in the gallery forest, linked to competition for light, could also explain the low diversity in this plant formation.

With regard to Shannon's diversity index, our results are close to those obtained by Djego & *al.* (2012) in the savannah formation of the Kaodji community forest in Benin, whose value is 2.33 bits.

In the partial reserve, the Shannon index varies from 2.40 to 2.99 bits depending on the plant formation: gallery forest (2.40 bits), tree savannah (2.61 bits), wooded savannah (2.63 bits), agroforestry park/ fallow (2.94 bits) and shrub savannah (2.99 bits). As in the total reserve, all plant formations have low diversity on the Frontier and Pichod-Viale scales, with the exception of shrub savannah, which has a diversity index of around 3 bits. Maximum diversity indices range from 3.09 bits (gallery forest) to 3.58 bits (shrub savannah and agroforestry park). The equitability index is average for wooded savannah (0.77), gallery forest (0.78) and wooded savannah (0.79); it is high for agroforestry park (0.82) and shrub savannah (0.83). On the Piélou scale, tree savannah, gallery forest and wooded savannah have an average species distribution, with indices of 0.77, 0.78 and 0.79 respectively. As for the agroforestry park/ fallow land and shrub savannah, the species present in these plant formations are almost equally distributed, with equitability indices of 0.82 and 0.83 respectively. Simpson's index ranges from 0.07 to 0.14. The values of this index indicate a low dominance of one or a few species in all the plant formations of the partial reserve. However, the shrub savannah is the plant unit in which species are almost equally distributed, with an index of 0.07. Ultimately, Simpson's inverse index confirms the diversity of the different plant formations in this reserve. Table 27 shows the distribution of diversity and equitability indices for plant formations in the partial reserve.

Table27 : Distribution of diversity and equitability indices for RPFB plant formations

Plant formations	Shannon diversity index (H′)	Maximum diversity (H'max)	Piélou(E) equitability index	Simpson Index (D)	Simpson's inverse index (1-D)
Gallery forest	2,40	3,09	0,78	0,14	0,86
Wooded savannah	2,63	3,33	0,79	0,12	0,88
Wooded savannah	2,61	3,40	0,77	0,10	0,90
Shrubby savannah	2,99	3,58	0,83	0,07	0,93
Agroforestry park/fallowland	2,94	3,58	0,82	0,11	0,89

Source: Floristic inventory data, December 2020

Analysis of the indices confirms that the shrub savannah is the most diverse and the gallery forest the least. All plant formations in the partial reserve are moderately diversified, compared with those in the total reserve, where the gallery forest and agroforestry/ fallow park are poorly diversified. In both reserves, the shrub savannah is the most diverse and the gallery forest the least. The former may be due to the regular presence of domestic animals in this vegetation unit. Anything that could lead to the dissemination and germination of various plant species by animals (zoochory). The soils on which shrub savannahs are planted are generally undeveloped soils on high glacis slopes; they are gravelly soils and armoured mounds. They are conducive to the development of numerous xerophilous species. The availability of light in savannah formations also favors the development of many plant species.

However, the low diversity of the gallery forest could be linked to the degradation of the riverbanks, in this case the Bougouriba, due to various anthropogenic pressures. Also, the numerical importance of a few species such as *Diospyros mespiliformis* (27.49%), *Anogeissus leiocarpus* (19.), *Balanites aegyptiaca* (12.) and *Mitragyna inermis* (8.) implies an environment conducive to the development of said plant species, or that these species have a propensity to colonize the space on which they grow. The edaphic conditions, distribution pattern and ecological plasticity of certain species strongly influence the density of individuals in the grouping (Mbayngone & *al.*, 2008).

4.3.4. Stand structure parameters

Structural parameters refer to all the elements used to evaluate the spatial (height) and horizontal growth of a plant stand.

4.3.4.1. Horizontal structure

Horizontal structure takes into account stand density, basal area and circumference class distribution.

4.3.4.1.1. Stand density and basal area

Observed density is one of the indicators used to characterize the dynamics of the horizontal structure of a plant stand. It varies with topography, soil, exposure, sylvigenetic stages, etc. (Pascal, 2003). The floristic inventory shows that the total reserve (245 individuals/ha) is denser than the partial reserve (159 individuals/ha).

The plant formations in the total reserve are classified in order of density as gallery forest (470 individuals/ha), wooded savannah (450 individuals/ha), wooded savannah (350 individuals/ha), shrub savannah (131 individuals/ha) and agroforestry park/ fallow land (104 individuals/ha).

As for the partial reserve, the plant formations in order of importance of woody density are: gallery forest (245 individuals/ha), wooded savannah (258 individuals/ha), wooded savannah (246 individuals/ha), shrub savannah (132 individuals/ha) and agroforestry park/ fallow land (90 individuals/ha). All plant densities are summarized in Table 28.

Table28 : Distribution of densities observed in reserve plant formations

Plant formations	Observed densities (individuals/ha)	
	Total reserve	Partial reserves
Gallery forest	470	245

Wooded savannah	450	258
Wooded savannah	350	246
Shrubby savannah	131	132
Agroforestry park / Fallow land	104	90
Reserves	245	159

Source: Floristic inventory data, December 2020

The table shows that almost all plant formations in the total reserve, with the exception of the shrub savannah, are denser than those in the partial reserve. The differences in density are very significant. They range from 220 individuals/ha for the gallery forest to 14 individuals/ha for the agroforestry/ fallow park.

The main reason for the differences in plant densities observed in the two reserves lies fundamentally in their status. In the total reserve, all human activity is prohibited, so protection measures are rigorous. On the other hand, as human presence is authorized in the partial reserve through the use of usufructuary rights and the presence of administrative villages within it, protection actions are often ineffective. Wood harvesting for various uses (firewood and timber) is very high, as is land clearing for agriculture, due to the geographical proximity and accessibility of the resource to local populations. This fact also justifies the lower densities observed in the shrub savannah and agroforestry/fallow park in both reserves. This finding corroborates the result of a study conducted by Diedhiou & *al.* (2018) in two villages in Senegal (Keur Birame and Saré Yorobana), which showed that the lowest actual vegetation densities are observed on cultivated units due to land clearing and poor farming practices.

Basal areas are low in almost all plant formations in both reserves, with the exception of the gallery forest in the total reserve , which has a basal area of 23 m²/ha; relatively close to the norm suggested by some authors for tropical dry forests. The lowest basal areas were recorded in the shrub savannah, with 3 and 2.76 m²/ha respectively in the total and partial reserves (see table 29).

Table29 : Distribution of land area in reserve plant formations

Plant formations	Land area (m²/ha)	
	Total reserve	Partial reserve
Gallery forest	23	10,45
Wooded savannah	8,91	9,53
Wooded savannah	8,87	6,55
Shrubby savannah	3	2,76
Agroforestry park/fallowland	6,26	6,18
Reserves	8,24	4,85

Source: Floristic inventory data, December 2020

Basal area is linked to stand density and average stem circumference. The denser the formation and the higher the average circumference, the greater the basal area. On the other hand, basal area is low in low-density stands with small stem circumferences. In this study, we found that the partial reserve had a low density (159 stems/ha) with a low basal area (4.85 m²/ha) compared to the total reserve (250 stems/ha) with a basal area of 8.24 m²/ha). These low basal areas are indicative of a stand dominated by young individuals or a clear stand. The results concur with those obtained in the Sota basin in northern Benin, characterized by a scarcity of large trees by Moumouni & *al.* (2019). Indeed, the similar ecological conditions of these two ecological environments located in the southern Sudanian domain may explain these results. This is the case for shrub savannahs dominated by stems of very low circumference, associated with low densities. On the other hand, in agroforestry/ fallow parks, densities are relatively lower than in shrub savannahs, even though they have larger land areas than the latter.

This is due to the relatively large number of large-circumference stems of species such as *Parkia biglobosa, Vitellaria paradoxa, Adansonia digitata and Faidherbia albida*, which are protected by the local population. The total reserve, which is better protected, has enough large-girth species such as *Anogeissus leiocarpus*, and a higher basal area than the partial reserve. This result confirms that obtained by Mbayngone & *al.* (2008) in the Pama reserve in south-east Burkina Faso. Both reserves are protected areas and are located in the same phytogeographical domain.

In view of these results, the management of these reserves for commercial timber exploitation would not be a relevant option in the sense that the exploitable timber potential is relatively low and would not promote good financial profitability, since basal area is, according to Zida & Ouadba (1998), the reflection of the stand's production capacity. Conversely, major reforestation of local species and protection of reserves would improve this index.

4.3.4.1.2. Distribution of woody plants by circumference class

4.3.4.1.2.1. Distribution of woody species by circumference class of total reserve

The distribution of woody plants by circumference class is unimodal for all plant formations on both reserves. Woody plants were grouped by circumference classes of 20 cm amplitude. In the total reserve, individuals are concentrated in small circumference classes in all plant formations. In addition, 35.64% of gallery forest trees are concentrated in the [30-50[class, 23.40% in the [50-70[class, 21.81% in the [70-90[class, 10.in the [90-110[class and 4.in the [110-130[class. Large circumference classes are very poorly represented in this unit.

Wooded savannah is represented by 60.74% of plant species in the [30-50[class, 29.in the [50-70[class and 8.in the [70-90[class. The others are almost non-existent.

In the wooded savannah, the circumference classes with the most trees are [30-50[, [50-70[and [70-90[, with 69.52%, 17.and 5.of the stand's individuals respectively. As in the wooded savannah, medium- and large-class individuals are very poorly represented.

The shrub savannah is occupied by 69.35%, 18., 5.and 4.of trees in circumference classes [30-50[, [50-70[, [70-90[and [90-110[respectively. Individuals belonging to large circumference classes are virtually non-existent.

Finally, in the agroforestry/ fallow park, 28.21% of individuals belong to the [30-50[class, 35.90% are concentrated in the [50-70[class, and 23.and 8.in the [70-90[and [90-110[classes respectively. On the other hand, the presence of individuals belonging to the ≥210 cm class was only observed in the gallery forest (1.06%) and in the agroforestry park (1.28%).

The statistics show a dominance of individuals belonging to the small circumference classes, notably [30-50[and [50-70[. The various vegetation formations in the total reserve are therefore covered with young trees, which would suggest that the reserve is in a regeneration phase. However, this observation may be explained by the negative influence of anthropogenic activities on the reserve's plant formations in the past, which would not have favoured the remarkable existence of old individuals with large circumferences today. In this respect, interpretation of the 1986 satellite image shows that 63.60% of the reserve's surface area was occupied by fields and fallow land. Protection measures taken by the State have encouraged the conversion of these formerly cultivated areas into shrub and tree savannahs.

4.3.4.1.2.2. Distribution of woody plants by circumference class of the partial reserve

In the partial reserve, the dominant classes are those of small circumference. In the gallery forest, the most representative classes are [30-50[, [50-70[, [70-90[and [90-110[, with 35., 30., 16.and 8.woody species respectively. These same classes are dominant in wooded savannah, with 46.63%, 22., 16.and 7.woody species respectively. Tree savannah is represented by 61.02%, 22., 8.and 5.of trees for classes [30-50[, [50-70[, [70-90[and [90-110[. As for the shrub savannah, three classes - [30-50[, [50-70[and [70-90[- contain 66., 22.and 7.stems respectively. Finally, the classes most representative of the agroforestry/fallow park are [30-50[, [50-70[, [70-90[, [90-110[and [110-130[, with 30., 21.52%, 17.94%, 11.and 9.87% stems respectively.

Statistics show that all plant formations in the partial reserve are occupied mainly by small-girth trees; large-girth trees are virtually non-existent in the tree and shrub savannahs, and are very poorly represented in the agroforestry/ fallow park. Like the total reserve, the plant formations of the partial reserve are therefore in a regeneration phase. This is due to the effects of agricultural activities through the clearing and slash-and-burn system. Referring to the 1986 satellite image, which is the oldest in the series of satellite images used, the total area of this reserve was occupied by 63.of fields and fallow land, which were subsequently converted to shrub savannah and then to tree savannah.

The distribution of woody plants by circumference class on both reserves is unimodal. Trees are highly concentrated in the small circumference classes, sparsely concentrated in the medium circumference classes and very sparsely present in the large circumference classes. However, the total reserve concentrates the highest frequency (54.83%) of ligneous trees belonging to the smallest circumference class [30-50[compared with the partial reserve, which has a frequency of 50.for the same class. For the other classes, which are fairly well represented, particularly those of [50-70[, [70-90[and [90-110[, the partial reserve is slightly dominant, with 23.versus 23.27%, 12.versus 12.and 6.versus 5.for the total reserve respectively (figure 11).

Figure11 : Circumference class distribution of RTFB and RPFB woody plants

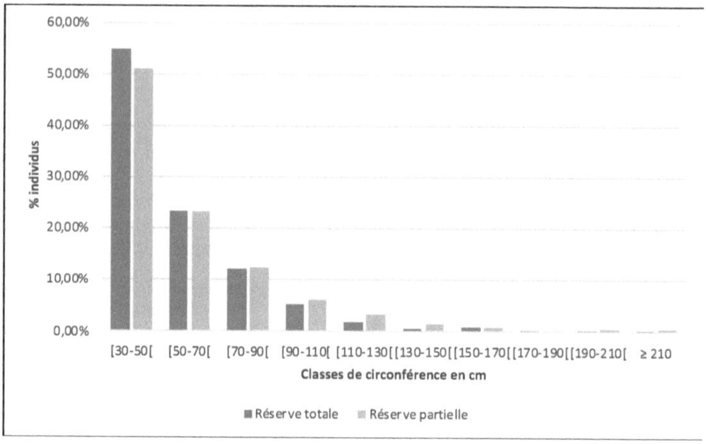

Source: Floristic inventory data, December 2020

Large-circumference trees are very rare in the plant formations of both reserves, whereas small-circumference young trees are numerically very important. The preponderance of young individuals, as mentioned above, could be linked to the slash-and-burn agricultural system. The same findings were made in the Tiogo classified forest in central-western Burkina Faso by Tankoano & *al.* (2016), and in the Nord-Borgou phytogeographical district in Benin by Moumouni & *al.* (2017), who established a relationship between cultivated clearings and the preponderance of low-girth individuals.

4.3.4.2. Vertical structure

Vertical structure refers to the organization of the woody stand into strata. It identifies trees from shrubs and saplings.

4.3.4.2.1. Height class distribution of plant formations in the total reserve

Distribution by height class shows a low proportion of individuals in the first class [2-4[for gallery forest (2.13%), wooded savannah (1.48%) and wooded savannah (7.). The same class is well represented in shrub savannah (23.39%) and agroforestry park/ fallow (11.). With the exception of this class, there is no great difference in the proportion of individuals between the other classes in gallery forest. Indeed, classes [4-6[, [6-8[, [8-10[, [10-12[, [12-14[, [14-16[and ≥16 are respectively represented by 9.57%, 15.43%, 14., 17.02%, 19.68%, 9.and 12.of individuals. On the other hand, in the wooded savannah, classes [4-6[, [6-8[, [8-10[and [10-12[are the most representative, with 9., 37., 30.and 15.of ligneous trees. The tall height classes are poorly represented in this plant formation, as well as in tree savannah, shrub savannah and agroforestry park/ fallow land. In the tree savannah, the dominant classes in terms of proportion of individuals are [4-6[, [6-8[and [8-10[for 30.47%, 37.and 17.of trees. As for the shrub savannah, classes [4-6[, [6-8[and [8-10[account for 45., 19.and 7.26% of individuals respectively.

In the agroforestry/fallow park, classes [2-4[, [4-6[, [6-8[, [8-10[, [10-12[and [12-14[are represented by 11.54%, 15.38%, 26., 19., 8.and 10.of woody plants.

Analysis of vertical structure allows us to characterize plant formations. Small woody plants (less than 7 metres tall) predominate in the shrub and tree savannahs. They are in the minority in gallery forest and wooded savannah. The gallery forest is the plant formation with the highest proportion of large trees.

4.3.4.2.2. Height class distribution of plant formations in the partial reserve

The distribution by height class shows relatively high proportions of small individuals in all plant formations in the partial reserve. Classes [2-4[and [4-6[respectively account for 7.and 21.05% of gallery forest trees, 11.and 21.of wooded savannah trees, 15.93% and 38.of tree savannah trees, 26.and 38.of shrub savannah trees, and 12.and 17.of agroforestry park trees. In classes [6-8[, [8-10[and [10-12[, 23., 23.and 15.of trees are found in gallery forest, and 29., 21.and 13.in wooded savannah, respectively, 26.10%, 13.and 4.for tree savannah, 21.43%, and 2.for shrub savannah, and 24.66%, 18.and 19.for agroforestry park/ fallow land. Class [12-14[is poorly represented in gallery forest (7.02%) and agroforestry/fallow park (6.). It is very poorly represented in wooded savannah (1.55%), wooded savannah (1.69%) and shrub savannah (0.). In the classes [14-16[and ≥ 16 m, there is virtually no proprotion of individuals in the wooded, tree and shrub savannahs, and very low rates in the gallery forest (0.58% and 1.78%) and the agroforestry/ fallow park (4.93%).

Examination of the above data reveals a significant presence of bushy individuals in all plant formations in the partial reserve. This is a sign of slowing vegetation growth. It can be explained by the strong animal and human pressure on the trees. Household surveys have shown that domestic animals from sedentary and transhumant stockbreeding are fed exclusively on crop residues and wild fodder. Regular grazing and trampling of trees by herds and other domestic animals, combined with wood cutting and late fire, therefore have a detrimental effect on the seasonal and even inter-annual growth of vegetation, creating shrub-dominated plant formations. The presence of bushy species in relatively high proportions is also linked to the conversion of fields into shrub savannah, which is scattered in small patches across the reserve.

4.3.4.2.3. Height class distribution of woody plants in the two reserves

If we look at the overall distribution by tree height class on both reserves, we can see that small and medium-sized trees predominate. In addition, classes [2-4[, [4-6[, [6-8[and [8-10[respectively account for 8.03%, 22., 27.and 17.of trees in the total reserve, and 16., 29., 24.and 16.30 % of trees in the partial reserve. In any case, and in view of the high proportions of small-circumference individuals mentioned above, the vegetation in these reserves would appear to have entered a regeneration phase, with a preponderance of young individuals. Figure 12 shows the distribution by height class of woody plants in the Bontioli total and partial wildlife reserves.

Figure12 : Height class distribution of RTFB and RPFB woody plants

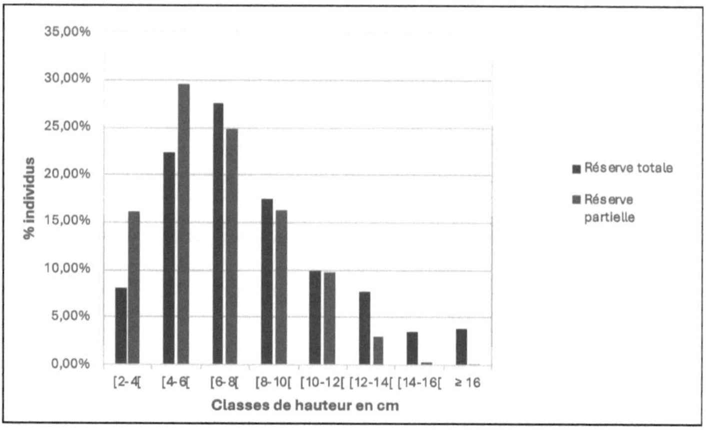

Source: Floristic inventory data, December 2020

Land use in the Bontioli wildlife reserves was marked by an increased human presence in 1986; more than half of these protected areas were occupied by fields and fallow land. The remaining units were spread over small portions. As a result of the protection measures launched by the environmental authorities, a substantial conversion of rain-fed crops to shrub savannah took place between 1986 and 2002. After this period, there was a sharp upsurge in the installation of fields between 2002 and 2018. All in all, the plant formations of these reserves are permanently disturbed by human activities.

Analysis of the characteristics of the plant formations reveals a relatively low floristic richness and diversity. Structural parameters indicate a preponderance of juvenile individuals, a sign that the future is possible if strong protection and conservation actions are implemented.

CHAPTER 5: THE EFFECTS OF CLIMATE VARIABILITY ON PLANT FORMATIONS

Climate change has captured the attention of scientific and political institutions the world over. Countries with fragile ecosystems, in this case those in the Sahel, are particularly vulnerable to the effects of climate variability on water resources, which are essential to the growth of vegetation and to the practice of many human activities in rural areas, notably agriculture and livestock breeding. The aim of this chapter is to identify the main climate parameters likely to vary, and to analyze their possible impact on the forest ecosystem of the Bontioli wildlife reserves.

5.1. Rainfall and the dynamics of reserve plant formations

Water is an essential factor in plant growth and reproduction. Analysis of the evolution of plant formations therefore requires prior knowledge of the spatial and temporal dynamics of rainfall.

5.1.1. Temporal dynamics of rainfall

Data from the Gaoua meteorological station, whose radius of action covers the Bontioli wildlife reserves and all their neighbouring villages, were considered due to the absence of a meteorological station in the study area. The rainfall trend curve for the Gaoua station in the series studied (1987-2017) (see figure 13), shows inter-annual irregularities through a sawtooth pattern. In 1991, the station recorded the highest water level in the series, i.e. 1435.6 mm, compared with 900.8 mm in 2007, corresponding to the lowest water level in the series. The average value for the series is 1081 mm of water. The trend line indicates a slight drop in rainfall in this zone. Thus, the drop in rainfall observed over the last thirty-one years of the series (1987-2017), endorses the overall decline in rainfall in West Africa. For, according to Allies (2018), this region of the globe is one of the most affected by current and projected climate change.

Figure13 : Interannual variation in rainfall (1987-2017)

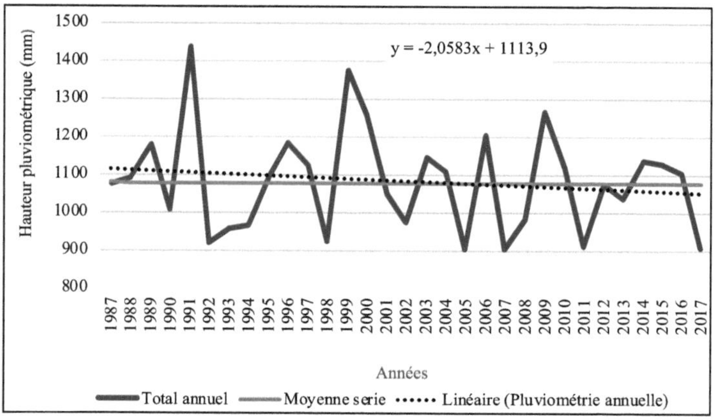

Source: Agence Nationale de la Météorologie, Gaoua station, 2019

These inter-annual irregularities in rainfall cause enormous difficulties in the ecological system and in agropastoral and forestry production. The success of the agricultural calendar by actors has become a lottery in a predominantly rain-fed cropping system. Because of its great temporal variability, rainfall is the main climatic factor influencing crop yields (Tinyar, 2003). Low yields due in part to rainfall irregularities prompt some farmers to clear stream banks for agricultural production.

5.1.2. Average monthly rainfall (1987-2017)

The umbro-thermal diagram (figure 14) shows that the rainy season starts in April and ends in October, i.e. six (6) months. During this period, the monsoon sweeps across the region, bringing rain. The wettest months are July, August and September, with a peak in August of 240 mm of rainfall on average. The least rainy months coincide with the beginning and end of the rainy season, i.e. April and October. Finally, the months of January, February, March, November and December are considered dry months, as water levels are lower than the respective average temperatures. This finding is in line with Henri Gaussen's theory, which states that a period is said to be arid when $P < 2T$. On the other hand, it is humid when $P > 2T$ (P represents the total amount of precipitation in millimetres for the given period, and T the average temperature in ° C for the same period).

The graph conceals numerous rainfall contrasts, insofar as pockets of drought are observed during the wet season. These are particularly hard on farming activities when they occur at the start of the season. Farmers are therefore obliged to sow the same fields two or three times. The early cessation of the rains reduces agricultural productivity, as the ears of millet, sorghum and maize are often seedless. The start of the rainy season in this zone is characterized by violent downpours, as is the case throughout the intertropical zone. These rains are therefore very aggressive, as they fall on bare, parched soils (Hauchart, 2005). They are therefore a source of water erosion, stripping the soil's superficial horizon and gullying it, exposing tree roots to the sun's rays and strong winds.

Furthermore, the length of the rainy season, the quantities of water precipitated and the rhythm of the rains are important indicators to consider when analyzing plant dynamics. This is a regional observation. In West Africa, the respective duration of the dry and wet seasons determines the type of vegetation and the quantity of biomass.

Average monthly temperatures vary only slightly. There are two cool periods: July to September and December to January. The lowest temperatures of the year occur in December, with an average of 25.7°C. By contrast, the highest temperatures are in April, with an average of 31.23° C. Finally, there are two hot periods: a post-wet period from October to November, and a pre-wet period from February to April.

Figure14 : Umbro-thermal diagram (1987-2017)

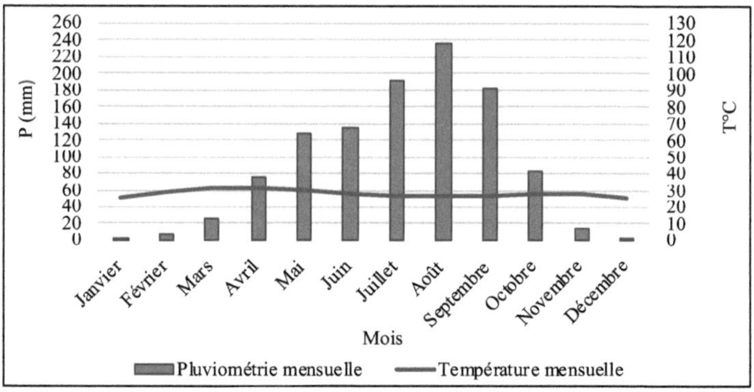

Source: Agence Nationale de la Météorologie, Gaoua station, 2019

5.1.3. ISP and determination of dry periods

Table 30 summarizes the different rainfall situations in the series (1987-2017) according to the ISP value. Figure 15 shows a rainfall trend characterized by wet and dry years.

Table30 : SPI value as a function of number of years

ISP value	ISP <-2	- 2<ISP <-1	- 1<ISP< 0	0<ISP< 1	1<IS P<2	ISP >2	Tot al
Years concerne d	-	1992 1998 2005 2007 2011 2017	1987 ; 1990 1993 ; 1994 2001; 2002 2008; 2012 2013	1998 ; 1989 1995; 1996 1997 ; 2003 2004; 2006 2010; 2014 2015 ; 2016	2000 2009	199 1 199 9	-
Number of years	0	6	9	12	2	2	31
Frequenc y (%)	0 %	19,35 %	29,03 %	38,71 %	6,45 %	6,45 %	100

Source: Agence Nationale de la Météorologie, Gaoua station, 2019

Of the 31 years in the series studied, 9 (29.03%) were years of moderate water stress, while 6 (19.) were years of severe drought. On the other hand, 38.71% or 12 years of the series are qualified as moderate humidity, while years of high and extreme humidity are respectively 6.(2 years) and 6.(2 years). However, over the whole series, a total of 16 wet years were recorded, representing 51.61%, and 15 dry years, or 48.. According to these statistics, rainfall in the area is relatively stable. The table shows the absence of any years of extreme drought and the presence of 2 years of extreme humidity in 1991 and 1999.

Nevertheless, in terms of evolution, the trend is towards a slight decrease in rainfall, as indicated by the trend line.

Figure15 : ISP trends (1987-2017)

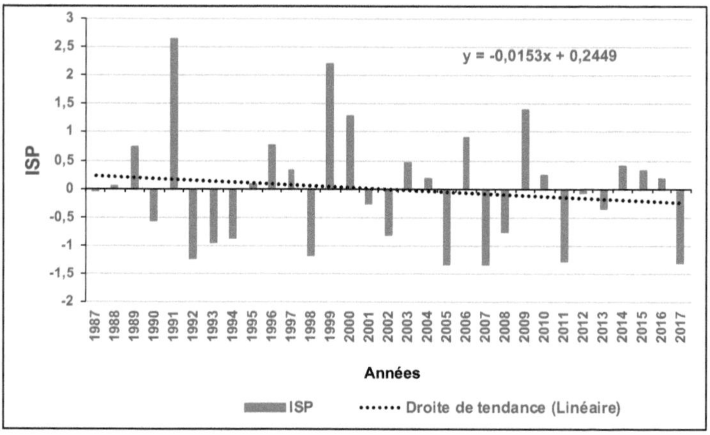

Source: Agence Nationale de la Météorologie, Gaoua station, 2019

5.1.4. Impact of the frequency of dry years on plant formations

An analysis of the graph helps us to understand the evolution of plant formations over this period. The impact of drought on vegetation in Sahelian countries is well documented. Indeed, according to Sarr (2008), persistent drought over several years causes ecological and economic imbalance. Water stress induces retraction of tree trunks, with a negative impact on stand productivity. This phenomenon was clearly described in the savannahs of western Burkina Faso by Devineau (1997). For him, high water stress is likely to influence the course of various vital processes. Trees undergoing prolonged water stress become dehydrated and their production is reduced.

Periods of water deficit are very often the result of increased clearing of riverbanks by farmers, the direct consequence of which is a reduction in the area of riparian formations and gallery forests. In the Bontioli wildlife reserves, gallery forests have been progressively replaced by fields over the period from 1986 to 2018. This situation is partly attributable to the drop in rainfall, which favours the clearing of land along riverbanks due to the lack of arable land, but also for its fertility. The development of plant formations depends not only on the amount of rainfall, but also on its temporal distribution (Koné & *al.*, 2007). A study carried out by Diello & *al.* (2005) in the Nakambé watershed in Burkina Faso supports the existence of a maximum of chlorophyll activity in plants at the heart of the rainy season, followed by a gradual decrease in vegetation index values from the end of the rainy season onwards, and lower and lower values as the dry season sets in. Elsewhere in Senegal, Solly & al (2020) have shown that variations in rainfall have led to tree mortality due to water stress, shortening of the vegetative cycle, reduced growth of certain species known for their height (*Khaya senegalensis*, *Ceiba pentandra*, *Afzelia africana*, and *pterocarpus erinaceus*) and scarcity of certain forest products. According to Traoré (2021), the rainfall deficit that has affected Sahelian countries since the late 1960s has had a lasting impact on key economic sectors and biodiversity.

On the other hand, Zorom & *al* (2022), in a study entitled "relationships between rainfall and vegetation production during the period 1999-2019 in Burkina Faso", showed a weak correlation between rainfall and vegetation index (FAPAR) in areas with a high presence of conservation areas and classified forests with a higher abundance of ligneous and other multi-annual species. The vegetation formations of the Bontioli wildlife reserves seem to obey this principle. In addition, despite the existence of years of severe and moderate drought over the series studied (1987-2017), there has been an increase in the area of certain plant formations, notably shrub savannah and wooded savannah. This can also be explained by the fact that the drop in rainfall is not very significant over the said period, but also that dry years are alternated with wet years; there has therefore been no long period of continuous water stress in this series that could have a significant negative impact on the evolution of these plant formations. Also, more than half of the years in the series are wet, while periods of moderate and high humidity are undoubtedly periods of plant growth.

5.2. Thermal characteristics of reserves

Temperature is one of the main components of climate. It influences the phenology of plant species through evaporation and transpiration, the means by which plants release some of the water they have accumulated into the atmosphere. The leaves of deciduous species are renewed in this way. Finally, the fruits of certain trees dry out under the action of heat.

5.2.1. Mean interannual maximum temperatures

The average maximum temperature curve (Figure 16) for the series shows very marked inter-annual variations. Its sawtooth shape is an indicator of this. 1998 is the warmest year in the series, with an average maximum temperature of 34.77°C. The lowest mean maximum temperature was recorded in 2012, with a value of 33.62° C. The average value for the series is 34.19° C.

Figure16 : Interannual variation in mean maximum temperatures (1987-2017)

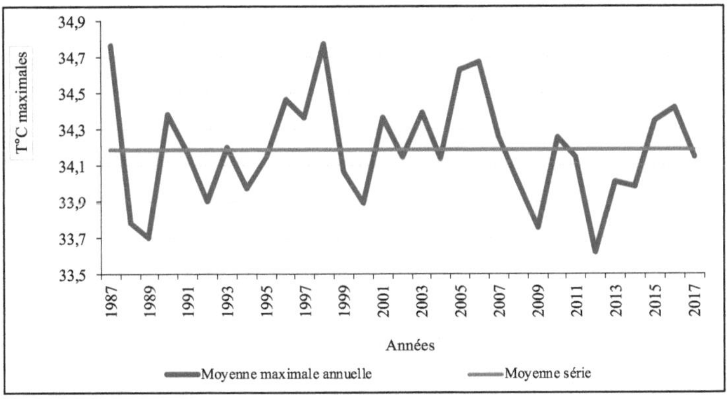

Source: Agence Nationale de la Météorologie, Gaoua station, 2019

5.2.2. Average interannual minimum temperatures

The inter-annual average minimum temperature curve (Figure 17) shows very wide variations from one year to the next. The highest mean minimum temperature was recorded in 2016, at 22.. On the other hand, the lowest mean minimum temperature was recorded in 2000 with a value of 20.833° C. The series average is 21.6° C. The trend line is rising. The progressive evolution of the minimum temperature implies a gradual warming of the area.

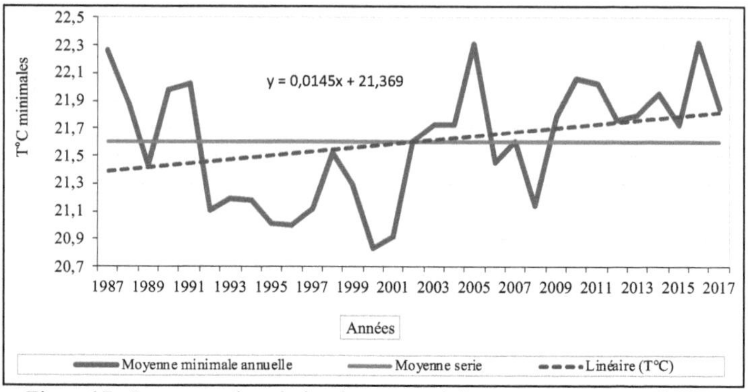

Figure17 : Interannual variation in mean minimum temperatures (1987-2017)

Source: Agence Nationale de la Météorologie, Gaoua station, 2019

5.2.3. Average monthly maximum and minimum temperatures

Figure 18 shows the curves for average monthly maximum and minimum temperatures in the same frame:

- November to February: this is the "coolest" period of the year, with minimum temperatures below . December is the month with the lowest average minimum temperature of the 1987-2017 series, at 16.61° C ;

- March to May is the hottest period of the year in the area. March saw the highest average maximum temperature (38.05°C) in the 1987-2017 series;

- June to September: wet, cool period with relatively low maximum and minimum temperatures;

- October to November: these two months coincide with the end of the rainy season and the start of the dry season. It's a post-wet, transitional phase, relatively warm.

It should be pointed out that average monthly temperatures (maximum and minimum) conceal many disparities, as there are extreme daily temperatures. During the cold season, some daily minimum temperatures can drop as low as 13° C or 14° C. Conversely, during hot spells, daily temperatures can approach 40° C. Hot spells are experienced during the cold season.

Figure18 : Trends in average monthly maximum and minimum temperatures (1987-2017)

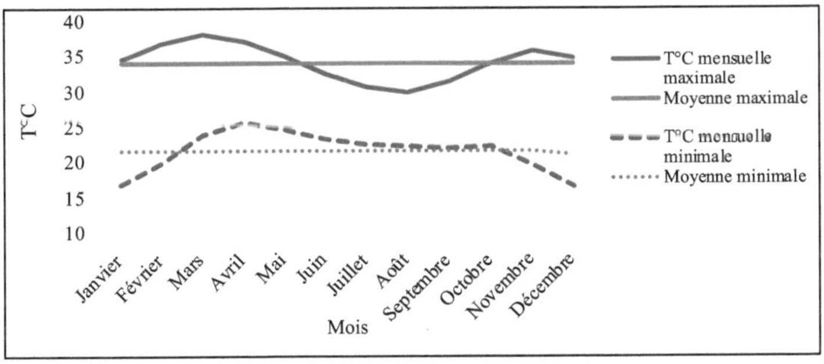

Source: Agence Nationale de la Météorologie, Gaoua station, 2019

5.2.4. Average monthly temperatures

Average monthly temperatures vary according to the time of year. The coldest month of the year is January, with an average temperature of 25.614°C. April is the warmest month of the year, with an average temperature of 32.232° C (figure 19).

Figure19 : Trends in average monthly temperatures (1987-2017)

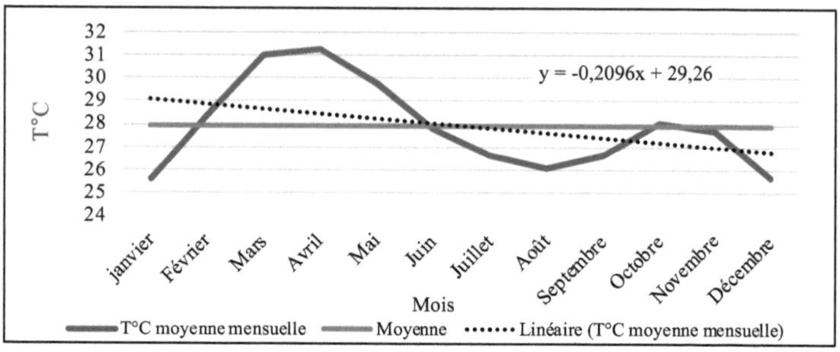

Source: Agence Nationale de la Météorologie, Gaoua station, 2019

5.2.5. Mean interannual temperatures

The mean annual temperature series (1987-2017) for the Gaoua station (see figure 20) shows interannual variation, as indicated by the sawtooth-shaped curve. The warmest and coldest years are 1987 (28.516° C) and 2000 (27.362° C) respectively. The series average is 27.897° C. However, the linear trend is upwards.

Figure20 : Interannual variation in mean temperatures (1987-2017)

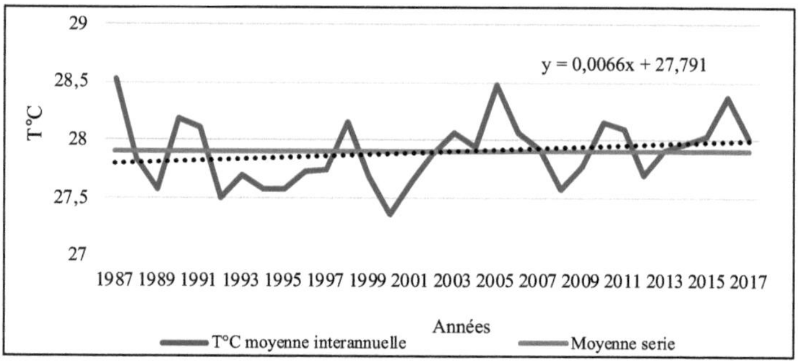

Source: Agence Nationale de la Météorologie, Gaoua station, 2019

5.6. Potential evapotranspiration trends

Potential evapotranspiration is a theoretical concept defined as the maximum quantity of water that could evaporate from a land surface for which water is not a limiting factor. It is opposed to actual evapotranspiration (AET), which is defined as the response of a given land surface to the demand exerted by ETP, given the water available to it (Allies, 2018). The determination of AET involves highly complex mechanisms.

In this study, it is the ETP that is considered, due to data availability. The curve has a general appearance of two quasi-symmetrical curves corresponding to the two main periods of annual ETP evolution. The period from February to June coincides with the hottest time of the year in the region, with high daily temperatures. PTE values are also high during this period (see figure 21). During this phase, the water table is lowered, as well as water levels in wells and most rivers dry up, leading to a reduction in plant activity. From October to February, many deciduous species in the reserves lose their leaves as a result of the drop in humidity caused by the drying effects of the harmattan. Leaf loss is a water-regulating mechanism used by these species. In addition, the ETP curve rises slightly during this period, when the average daily and monthly temperatures recorded in the locality are relatively low. From July to September, the period of low monthly temperatures, the area records its lowest ETP level. Plant growth is at its peak.

Figure21 : Monthly trends in ETP (1987-2017)

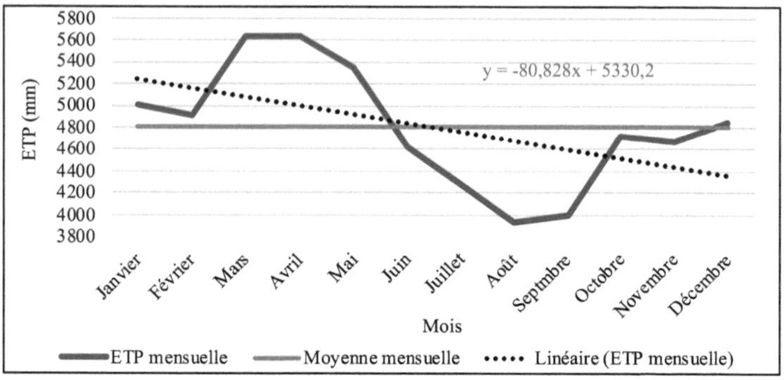

Source: Agence Nationale de la Météorologie, Gaoua station, 2019

5.2.7. Impact of temperature on plant formations

The rise in temperature in the study area, located in the Sudanian domain, confirms the findings of the IPCC (2007) regarding the gradual global rise in the Earth's temperature.

Rising inter-annual temperatures inevitably lead to a rise in inter-annual PTE, with adverse consequences for the availability of water resources (surface water and groundwater). While severe and prolonged water stress can lead to the death of plant species.

It should be noted that the rise in atmospheric temperature accentuates plant evapotranspiration; this has a negative influence on plant growth through reduced CO_2 uptake at- above a certain diurnal temperature threshold, which is 32° C for tropical forests .[17]

In the Bontioli reserves, the phenomenon of tree mortality is not very noticeable, but could be concealed by the logging activities that regularly take place in these protected areas. During the focus groups,

[17] lesechos.fr/monde/enjeux; Muryel Jacque, 2020 expires on 21/06/2023

the women stated that their priority was to cut dead wood. Nevertheless, according to the samples inventoried, standing dead wood was found in the agroforestry park/ fallow forest (8.87%) and gallery forest (0.74%) of the total reserve. In the partial reserve, 0.58% and 0.45% of standing dead trees were found in the gallery forest and agroforestry park/ fallow land respectively.

The work of Adjonou & *al* (2009), carried out in the open forests of the Oti-Kéran National Park in North Togo under a Sudanian climate, showed that rising temperatures and falling rainfall would cause the climate to dry out and reduce the availability of water and nutrients, limiting the physiological functioning of trees, especially those of small diameter. In this study, the phenomenon of tree mortality was attributed to climate change, in particular rising temperatures and falling rainfall in this region.

Furthermore, a study conducted by Oloukoi & *al.* (2014) in Central Benin showed the existence of a positively significant correlation between maximum temperature and the radiative power of fires. In other words, an increase in temperature enhances the radiative power of fires. The severity of fires can lead to the destruction of the most sensitive species in plant formations. These changes in thermal and hydric conditions have an impact on the frequency of forest fires (Da, 1996).

5.4. Winds

The study area is mainly under the influence of two main trade winds, the monsoon and the harmattan, whose speed variations and impacts on plant formations are analyzed in this section.

5.4.1. Average monthly wind speed

The average monthly wind speed curve for the 1987-2017 series (see Figure 22) rises from November to April, when it peaks at an average speed of 2.43 m/s, corresponding to the start of the rainy season characterized by showers. It then begins to decrease, reaching its lowest level in November (0.94 m/s). The average of the series is 1.6 m/s.

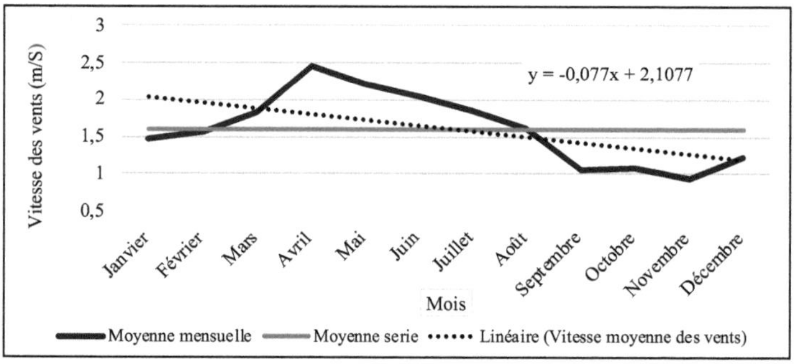

Figure22 : Monthly average wind speed (1987-2017)

Source: Agence Nationale de la Météorologie, Gaoua station, 2019

5.4.2. Interannual mean speed

The interannual mean wind speed curve for the 1987-2017 series at the Gaoua station (Figure 23) shows a generally increasing trend. The trend line is also increasing. The years with the lowest and highest wind speeds are 1987 (1.016 m/s), 2008 and 2010 respectively, when wind speeds reached 1.916 m/s. The area is thus characterized by a gradual and fairly marked increase in wind speed, as indicated by the trend line.

Figure23 : Variation in interannual mean wind speed (1987-2017)

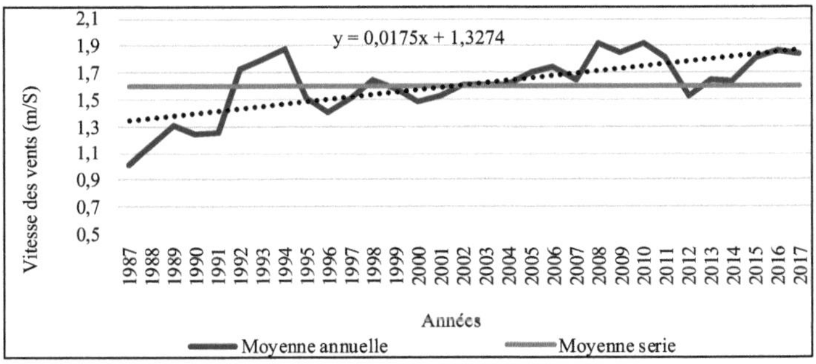

Source: Agence Nationale de la Météorologie, Gaoua station, 2019

5.4.3. Impact of wind speed on reserve plant formations

Winds have an adverse effect on plant formations, with some trees uprooted and flowers and fruit destroyed by strong winds, especially at the start of the rainy season. They also cause the exudation of certain plant species. Unfortunately, this period coincides with the fruiting and ripening phase of *Vitellaria paradoxa*, a plant species that is very present in the reserves and in all their villages, part of whose fruit is destroyed by the action of the winds.

The speed of fire propagation is also a function of wind speed. Due to the availability and quantity of herbaceous biomass in reserves, harmattan accelerates the intensity of fires, especially when they are set in the afternoons, when they gain in intensity; large areas of forest are thus often devastated by fire. Seasonal bushfires influence the resilience mechanisms of plants in these protected areas. Harmattan is therefore an important factor in the phenological cycle of Sudanian plant species. A study by Mahamane & *al* (2007) in Niger's "W" park, under a northern Sudanian climate, concludes that the various phenological stages are correlated with climatic parameters, including temperature.

Sand wind influences the phenological functioning of plants. Chlorophyll activity is reduced by the closure of plant stomata, resulting in the drying out of all or part of the tree. This phenomenon causes plant stands to die back. Photo 2 shows plants uprooted by the wind in the ve total fauna reserve at Bontioli.

Photo2 : Plants uprooted by the wind in RTFB

Field shot (SOME T. Noël, December 21, 2020)

Wind also plays a part in the dissemination of plant seeds and pollen. This mechanism, known as anemochory, contributes to the regeneration of plant formations in conditions where young shoots are preserved from the harmful effects of fire and grazing.

5.5. Sun exposure

Sun exposure refers to the duration of sunlight. It is expressed in hours.

5.5.1. Average monthly insolation

Figure 24 shows the variation in average monthly insolation for the 1987-2017 series. The brightest month is December (276.04 hours), while the month with the lowest insolation is August (173.25 hours). This observation was made in the Benin basin of the Niger River by Vissin (2007). According to Oloukoi & *al.* (2014), the longer the duration of sunshine, the more violent the vegetation fire. August being the wettest month of the year in the study area during which cloud cover is at its maximum stage, solar rays do not reach the earth's surface optimally. The monthly average is 239.218 hours. Sunlight is necessary for the flowering of certain trees, notably *Vitellaria paradoxa* (shea), as mentioned by Guira (1997). Plant reproduction processes that could be affected by climatic factors include flowering, pollination, seed production and germination (Chidumayo, & *al.,* 2011). Given the importance of insolation in the area, solar energy could be used as a substitute for wood energy and, as Mainguet (1995) argues, in dry ecosystems, solar energy should be a priority energy source over trees.

Figure24 : Average monthly insolation (1987-2017)

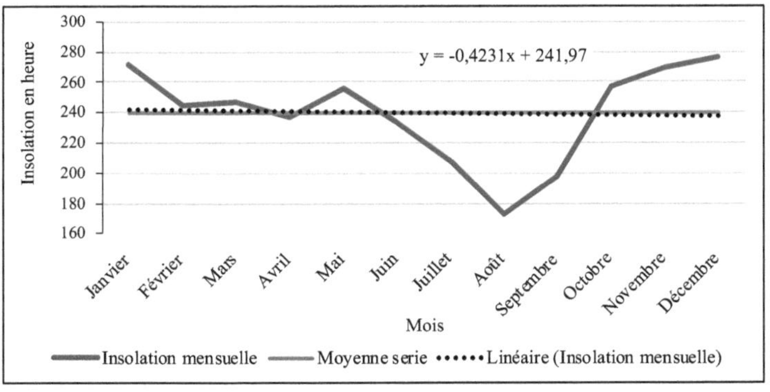

Source: Agence Nationale de la Météorologie, Gaoua station, 2019

5.5.2. Average interannual insolation

The curve (figure 25) shows very marked variations in the number of hours of sunshine over the years. For example, 1990 was the year in which the sun shone most brightly over the entire series (3015.7 hours), whereas it shone least brightly in 2004 (2689.1 hours). The average for the series is 2871.958 hours. The linear insolation trend for the series is decreasing.

Figure25 : Interannual variation in insolation (1987-2017)

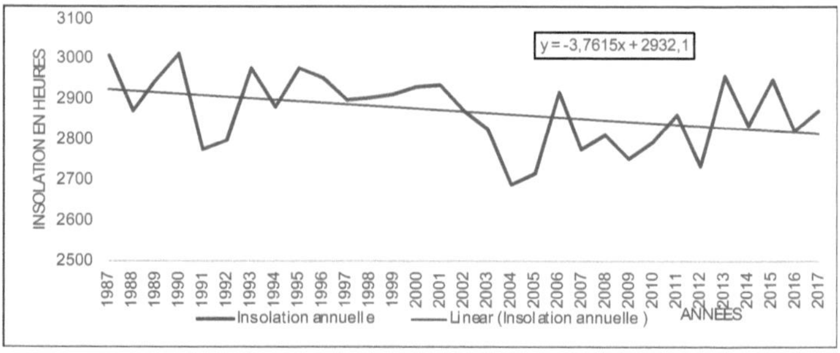

Source: Agence Nationale de la Météorologie, Gaoua station, 2019

Climate certainly has an influence on the ecological system. The main climatic parameters analyzed in this chapter, i.e. rainfall, temperature, ETP, wind speed and insolation, indicate overall climatic variability in the area, characterized by a downward trend in all the above-mentioned climatic parameters over the period 1987-2017. Nevertheless, the downward trend in rainfall over the said period, manifested by years of severe and moderate drought, associated with the rise in temperature as well as PTE and wind speed over the same series have not significantly negatively influenced plant formations, even if some direct negative impacts have been reported, in this case wind (speed and nature) on the phenology of certain plant species, and indirect ones such as the clearing of riverbanks for agricultural activities following periods of drought.

Partial conclusion

Land-use dynamics in the two reserves are marked by a jagged evolution of units. There have been both gains and losses of units in the various vegetation formations. However, it is the fields and fallows, and the shrub savannah that have seen the most change. The first unit has regressed significantly, while the second has progressed considerably.

Analysis of vegetation parameters indicates low diversity and richness of flora in almost all plant formations on both reseves. Nevertheless, the predominance of young stems could encourage good regeneration if protection measures are stepped up.

The biophysical and human characteristics of the Bontioli wildlife reserves were analyzed. The geographical position of these reserves, located in the southern Sudanian domain, is an advantage for the development of flora. The analysis carried out in this section also shows that the area's climatic parameters, like those of Burkina Faso, are deteriorating through an increase in temperature, potential evapotranspiration, wind speed and a reduction in rainfall over the period 1987-2017. However, the deterioration in the climatic parameters studied has not had any perceptible negative influence on the evolution of plant formations in these reserves.

PART THREE: MANAGEMENT METHODS AND THE VULNERABILITY OF PLANT FORMATIONS ON RESERVES

It consists of two chapters:

- Chapter 6: The impact of agricultural and pastoral practices on the evolution of plant formations in the reserves. It analyzes the agrarian systems of farmers in villages bordering the reserves, through their agricultural and pastoral production systems and the various repercussions on plant formations.

- Chapter 7: Reserve management and protection. This chapter takes stock of the evolution of management methods applied in the reserves, and the various implications they give rise to. It also analyzes the socio-economic spin-offs from the exploitation of forest products and the conditions of application of the Forestry Code.

CHAPTER 6: THE IMPACT OF FARMING AND GRAZING PRACTICES ON THE EVOLUTION OF PLANT FORMATIONS

In the quest for survival, mankind has always relied on nature to produce, feed, heal or provide shelter. The balance of ecosystems was maintained insofar as harvesting was either less than or equal to their regenerative capacity. In view of the demographic boom triggered by the improvement in the health conditions of African populations, particularly those of Burkina Faso, since the end of the twentieth century, and the introduction and development of production geared exclusively to the market, in this case cotton, forest ecosystems as a whole are now under unprecedented pressure. In this chapter, we analyse agrarian systems through agricultural and pastoral production systems, the various techniques involved and their repercussions on plant formations.

6.1. The agricultural production system

Land management in terms of access, security and development is a key factor in any agricultural production system.

6.1.1. A two-faceted rural land tenure system

Rural land tenure has its origins and foundations in the socio-political and religious organization, as well as in the rules of law that govern the various societies. Two land tenure systems currently coexist in the villages bordering the Bontioli wildlife reserves. One is based on traditional customary law, according to which land management is based on social relations based on kinship ties. This system is analyzed in terms of the legitimacy of land ownership, which should benefit the descendants of landowners. Traditional land law is based on the principle that there is a natural right to land for every human being, the right of every human being, whether or not he or she belongs to the community, to be allocated enough land for his or her subsistence and that of his or her family (Kaboré, 2010). The other is based on the legality that every citizen should have over the land. It is based on legislative texts and laws.

6.1.1.1. Traditional land tenure

In many societies in the south-west, land is family or clan property. Ownership is therefore collective (Dialla, 2003). It is an asset intended for agricultural production and is generally passed on by bequest, from father to adult son (a boy who has founded a family through marriage in the traditional context), by permanent gift to nephews and cousins close to the family, or by temporary gift to any person requesting land for agricultural production. In rural areas, and particularly in Burkina Faso, membership of local communities is a determining factor. Settlement history and seniority play a key role. The kinship system, which translates into membership of a lineage, is increasingly important for access to land and resources. Depending on whether the individual belongs to the lineage of the village founder, he or she will have privileged access to land and resources, as the ancestor is the one who first occupied the site and made a pact with the tutelary gods (Zonou, 2008).

In the villages bordering the Bontioli wildlife reserves, land is a non-market good. Access to it for farming is governed by very specific rules.

According to the people interviewed, 85.8% obtained their farmland by bequest, 16.by gift and 1.by loan. It is clear, therefore, that the main method of land management is still associated with customary law. It is passed down through the family.

In principle, in the traditional land tenure system, land cannot be sold and should remain the exclusive property of the holding family under the control of the land chief, who is a religious authority in charge of traditional land issues. Since land is a family asset that is passed down through the generations, a woman does not inherit land in the same way as an unmarried son, but the latter has the privilege of remaining under his father's guardianship until his death.

Migrant farmers' access to land is governed by religious and cultic norms. Applicants for agricultural space contact the customary authority. The latter will advise them on how to obtain a plot of land for agricultural production. Customary practices differ from one locality to another. Generally speaking, however, livestock and poultry (goats, sheep, chickens, guinea fowl, etc.) are required of those applying for agricultural land. The fields granted to migrants are former fallow land or bush (savannah). At the end of the sacrifices, which authorize the newcomer to carry out any activity on the land (building a dwelling house, opening a field), principles are laid down for the newcomer, and failure to respect them can result in expropriation. This is what Zougouri (2006) calls "religious clauses", which are explicit about what the beneficiary must do to respect the land on which he or she has settled. They are not allowed to plant trees in the fields. This is almost a general observation in areas where migrant farmers have settled. This is perceived by landowners as appropriation of the farm plots

(Gomgnimbou & *al.,* 2010). The same observation was made in agrosytems in Tem country in Togo by Koumoi & Laré (2014). They argue that the act of planting a tree in areas of recent colonization is perceived as a sign of appropriation of the land by the land chief. These practices on the part of land chiefs are therefore not conducive to improving the productive capacity of non-natives' fields through reforestation. Some migrants are victims of the saturation of agricultural space. Without necessarily committing a fault, they are often evicted from their land by their heirs.

The traditional land tenure system in villages bordering reserves does not guarantee security for producers, especially "foreign" producers, and hence for agricultural production in the locality. As a result, new fields are being created in the reserves. Customary leaders, in violation of forest legislation, grant fields to agricultural migrants in the reserves. According to the CILSS (2002), the Sudanian zones are currently under considerable pressure, as they still contain large land reserves. New arrivals are conquering areas previously given over to shrub savannah and forest, and bringing them under cultivation after clearing and slash-and-burn.

6.1.1.2. Modern land tenure

Since the advent of the People's Democratic Revolution in 1984, the Agrarian and Land Reform (RAF) has been adopted, bringing a new dynamic to land management. It advocates equal access to land for men and women, and proclaims land as a State asset. Article 4 of this law stipulates that "the national land domain is the property of the State as of right" (Bouda, 2008). The RAF was revised in 1994, 1996 and again in 2009, resulting in Law N°0034-2009/AN on rural land tenure in Burkina Faso. It reaffirms that land belongs exclusively to the State and its decentralized authorities. However, it has to be said that the implementation of this land legislation is far from being a reality, insofar as the interests of rural stakeholders are very divergent. As land chiefs are not sufficiently aware of the importance and scope of this legal provision, their authority is being eroded.

6.1.1.3. Land conflicts

Unclear land management is a source of insecurity and does not encourage optimal, sustainable use of natural resources. Competition between stakeholders in the exploitation of natural resources gives rise to pronounced differences that very often degenerate into conflicts with unfortunate consequences. The most recurrent disputes are those between farmers and herders, and between farmers and farmers. Transhumance corridors exist in the locality. However, they are obstructed in places by fields and gardens, forcing transhumant pastoralists to use the reserves during their seasonal movements, according to the Eaux et Forêts agent stationed in Zambo. This situation inevitably leads to conflicts when livestock destroy fields. Disputes between farmers arise in situations where the boundary between two fields is not clearly marked, or where the rightful owner attempts to expropriate the occupier of the land. These conflicts often become family, clan or community-based. The words of an interlocutor from the village of Bontioli clearly confirm the existence of this type of discord "(...) we had a conflict with the people from the neighboring village over a piece of land they wanted to take away from us. We decided to swear on customs to identify its true owners (...). The situation ended with the death of their customary chief. The corollary of this conflict was a massive clearing (dozens of hectares) of woodland in the total reserve in retaliation.

6.1.2. The persistence of a traditional farming system

Traditional agriculture is mainly aimed at satisfying family food needs. Farmers therefore rely on rudimentary farming techniques and equipment. In many African countries, farming systems remain unsustainable and insufficiently intensified, with low yields (Abba & *al.,* 2006).

6.1.2.1. Traditional farming techniques and equipment

Traditional cultivation techniques involve opening up a new field to be ploughed at the end of the rainy season (September- October) using the daba. At the end of the dry season, shrubs and certain trees that are not useful or considered troublesome are cut or pruned and piled up to be burned (photo 3). The newly-opened field is ready for sowing. The management of old fields relies on the management of harvest residues prior to sowing. Stems, shrubs and perennial grasses are dug up with dabas and cutters and burned. Ploughing, weeding, hoeing and ridging are carried out manually using rudimentary tools. The main soil conservation technique in bush fields is based on the maintenance of so-called utility trees. Some farmers, still using traditional cultivation techniques, are gradually adding stone cordons (6.of those surveyed) to fertilize their fields. The vast majority of farmers practice *mining-type* agriculture in bush fields, which receive no organic or mineral fertilizer, the corollary of which is a decline in soil fertility. Kadjo (2010), attests that these extensive cultivation practices and the low level of technicality of agriculture, has led to excessive consumption of resources and their degradation, in the absence of measures to protect or restore the soil capital in the commune of Dano. In addition, the hut fields benefit from household waste and animal manure.

Photo3 : Clearing a new field in the RPFB

Field shot (SOME T. Noël, March 23, 2020)

This traditional form of rural land management is now being severely challenged by population growth and the emergence of space-consuming cash crops, which have led to the saturation of rural areas. The application of modern techniques has enabled the production of crops traditionally reserved for hut fields, such as maize in bush fields.

In addition, the slash-and-burn technique has consequences for the regeneration of shrub-savannah plant formations, since it is the bushy plant species that are destroyed during clearing by fire and cutting. This technique could also have a negative impact on biodiversity, in that the selective practice of maintaining species in the fields could drastically reduce or even eliminate the populations of certain "non-utilitarian" plant species, especially bushy species, in the reserve ecosystem. This technique constitutes the greatest threat to the natural formations and biological diversity (Sawadogo, 2009) of these ecosystems. Selective felling of trees and shrubs followed by ploughing is practised by 96.of respondents who have opened new fields. This agricultural practice was observed in the Bondoukuy savannahs of western Burkina Faso by Devineau (1998). For this author, when clearing land, useful species (mainly shea, but also several others) are retained, all others being eliminated. This traditional system of cultivation is seen as the cause of deforestation, insofar as bare land increases due to the low level of vegetation regeneration resulting from the regular development of cleared areas. When these clearings are combined with animal-drawn ploughing, water erosion accelerates, and soil loss is even greater than in fields where clearing is followed by daba ploughing, then sowing. The practice of direct seeding under litter has proved effective in reducing runoff and erosion, including carbon losses (Palé, 2020).

6.1.2.2. Fallowing as a traditional soil fertility management method

Fallow land, an operative concept, is now studied by agronomists and geographers as a breeding ground for flora and fauna, and as an integral element of an agricultural landscape, used by man (Fourgeau, 2002). This is a traditional soil fertility management technique used in slash-and-burn farming systems, where soils do not receive the external input of organic or mineral matter essential to their regeneration. The reconstitution of fertility depends on that of herbaceous and plant biomass. Fallowing is still practiced by 45.50% of respondents, compared with 54.50% who no longer do so. These statistics show that fallowing has been abandoned by a significant number of farmers due to the saturation of rural space. Those who have the privilege of using this soil fertility management technique do so at the expense of reserve plant formations. New fields are inevitably cleared in these plant formations, as there is no other bush for the villages located within the reserves. This act is carried out in the forest by 10.50% of respondents, who are in fact landowners and migrants.

The length of time a field is left fallow after several years of cultivation is very short in the study area. Surveys reveal that 24.60%, 15., and 1.20% of respondents leave their fields to rest between 1 and 3 years; 4 and 5 years; 6 and 10 years; 11 and 15 years respectively, after several years of cultivation (see figure 26). This situation limits the regeneration potential of protective soil vegetation (Bandré, 1996).

Figure26 : Fallow periodicity

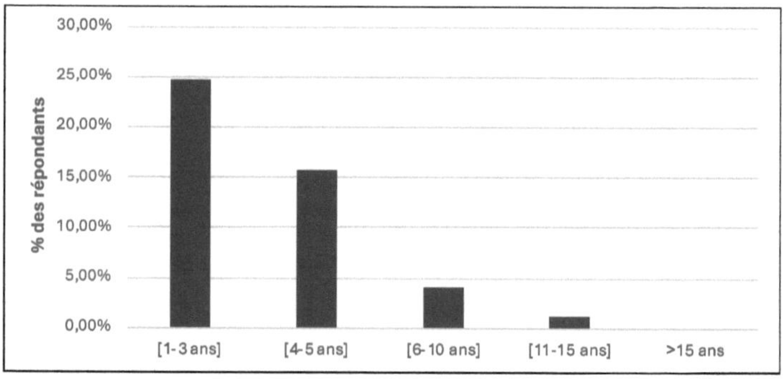

Source: Field data, March 2020

Analysis of the graph shows that land in the villages bordering the reserves is constantly being developed. According to 87.4% of respondents, this is due to the lack of agricultural land, compared with

7.who believe it is due to the increase in agricultural land. Increasing production needs for the cotton market and the saturation of agricultural land due to demographic pressure are the main reasons for the reduction in fallow periods.

6.1.2.3. Areas sown and crops produced

Assessing the areas sown by farmers is a very difficult task, given the increasing illiteracy of the main players. Nevertheless, class intervals have been identified in which the average areas sown by farmers are located. In fact, most of these fields are cleared in the partial reserve, which includes 10 villages from the survey sites, and a very small proportion is sown in the total reserve. Table 31 shows the areas sown by class intervals.

Table31 : Areas planted

Field areas	Workforce	Frequency (%)
Less than 1 ha	7	2,20
1 ha-3 ha] [1 ha-3 ha	129	39,37
[4 ha-6 ha]	123	37,80
[7 ha-10 ha].	50	15,40
[11 ha-15 ha]	13	4,00
[16 ha-20 ha].	2	0,60
More than 20 ha	1	0,30
Total	325	100

Source: Field data, March 2020

The table shows that 39.and 37.of farmers surveyed sow fields in the ranges [1 ha-3 ha] and [4 ha-6 ha] respectively; only 15.4% and 4.of respondents clear fields in the ranges [7 ha-15 ha] and [11 ha-15 ha]. These small areas intended for agricultural production indicate a real saturation of space. In fact, 64.6% of those interviewed stated that they no longer had the space to open new fields. The farming tools and techniques used are the foundation of this type of cultivation system, which is fundamentally oriented towards household consumption. According to our investigations, each head of household has an average of six (6) mouths to feed.

A variety of farm machinery is used in agricultural production. These include the daba, the animal-drawn plough, the tractor and other equipment (cutters, etc.), used respectively by 76.4%, , 3.and 0.of respondents (figure 27).

Figure27 : Means of agricultural production

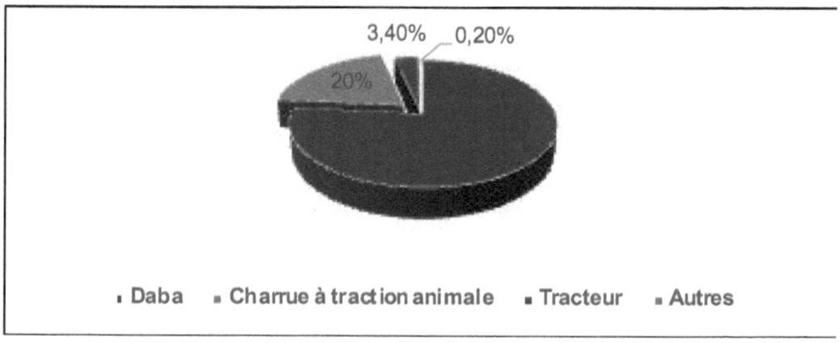

Source: Field data, March 2020

The graph shows that the most widely used agricultural production tool is the daba. This is characteristic of an unsophisticated form of agriculture. However, this type of farming has begun to modernize, with the introduction of animal-drawn ploughs and tractors.

The low level of agricultural equipment available to farmers in the study area leads them to adopt expansive farming practices based essentially on slash-and-burn. These farming practices in the reserves lead to a fragmentation of their plant formations, with the most wooded evolving into units that are poor in diversity and population, notably fallow land.

Nevertheless, cotton production has increased in the locality, making it the leading cash crop.

6.1.3. Cash crops in the agricultural production system and their impact on reserve plant formations

Cash crops were introduced in Upper Volta in a very deleterious socio-political, economic and environmental context.

6.1.3.1. The use of innovative techniques and methods by players in cash crop production

Cash crops benefit from new farming techniques and innovative methods, for reasons mainly linked to the profit expected by producers to pay off their debts. The row-seeding method used by cotton growers encourages the use of semi-mechanized farming equipment such as animal-drawn ploughs for weeding. The use of chemical fertilizers and pesticide treatment of the cotton plant, combined with a high level of labor input, increases yields.

Soil fertilization techniques are not very diversified in the zone. Indeed, 59.1% of farmers surveyed claimed to use chemical fertilizers (NPK and urea) to fertilize the soil, compared with 42.who used organic manure. Stone cordons and crop rotation are soil conservation and regeneration techniques that are rarely used in the locality. Only 6.8% and 3.of respondents said they used these techniques (figure 28).

Figure28 : Soil fertilization techniques

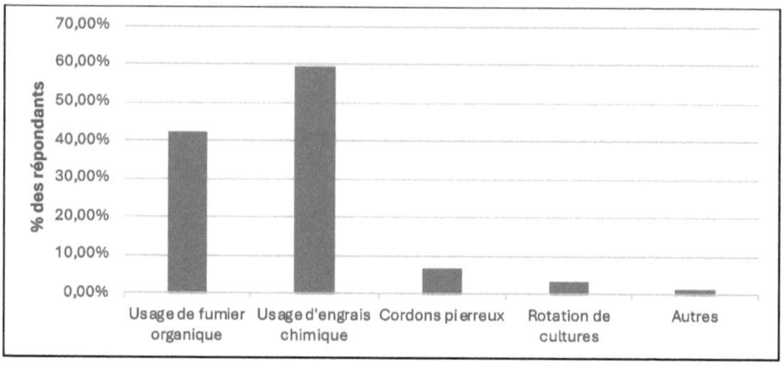

Source: Field data, March 2020

The use of chemical and organic fertilizers are the practices most frequently employed by farmers. Cotton growers strive to implement innovative farming techniques advised by agricultural agents. Agricultural inputs (fertilizers, pesticides, etc.), obtained from cotton production groups under the aegis of the Société des Fibres Textiles (SOFITEX) in the form of credit, payable during marketing, are used to fertilize cotton fields and treat the cotton plant against any attack by harmful insects. Manure pits are generally set up by farmers with a herd of oxen or draught oxen. Animal droppings, crop residues and household waste are dumped into these pits for decomposition. At the start of the rainy season, this waste, decomposed into organic matter, is spread on the fields before ploughing. Stone barriers are used to slow down the effects of water erosion. Boulders are aligned perpendicular to the direction of the slope. Finally, those who use crop rotation do so to enable the crops to benefit from the effects of chemical inputs (NPK and urea) from the previous season's cotton fields. It also aims to maintain soil fertility.

The use of chemical fertilizers in the agricultural production system has a dual effect reserve plant formations. In the short term, chemical inputs improve soil productivity, resulting in higher crop yields, especially for cotton. In the long term, however, they lead to chemical degradation of the soil through acidification (R. Lokossou & *al.*, 2018), with adverse effects on crop yields. Farmers will therefore be obliged to regularly increase the quantities of chemical inputs used in the fields or resort to organic inputs to maintain production levels at a satisfactory threshold. If need be, they will have clear the vegetation in the reserves.

6.1.3.2. Cotton growing at the heart of environmental issues

Cotton growing has gradually become the main cash crop in the South-West region, and more specifically in the Ioba and Bougouriba provinces, since the liberation of this region and the establishment of farming communities under the Autorité de l'Aménagement des Vallées des Voltas (AAVV) program set up in 1974. This state of affairs led to a significant increase in rural land requirements, as the high rate of growth in cotton production in the area was due to the increase in the area sown to this crop. As a result, fallow land tends to disappear; when it does exist, its duration is almost insignificant in terms of promoting real regeneration of soil suitable for agriculture. This is despite the fact that fallowing is one of the best practices for conserving soil fertility and maintaining vegetation (Somé, 2016) in the rural agrarian systems of Sahelian Africa, where trees play a vital role in agricultural production.

Over the last fifteen years, cotton production in the three provinces of the South-West region bordering the Bontioli wildlife reserves -

Bougouriba, Ioba and Poni - has grown relatively significantly. This is the result of an increase in the area sown to this crop (figure 29). A clear indicator between the rate of occupation of agricultural land and the expansion of cotton growing is the increased encroachment on protected areas for farming in recent years (Kaboré, 2010). The province of Noumbiel has been downgraded in the analysis for the simple reason that over several years it has not recorded any cotton cultivation, and it also does not border the wildlife reserves that constitute the site of the present research. The graphs below clearly show the predominance of cotton cultivation in the province of Ioba over the other two; a province in which a large proportion of the reserves are located.

Overall, there is an upward trend in cotton acreage in all three provinces. However, the province of Ioba is recording greater growth in cotton-growing areas than the other two provinces bordering the reserves. New fields are being opened up in former fallow land, protected and classified forests and reserves, and/or fields intended for cereal crops are being converted for cotton production.

Figure29 : Change in area sown (in ha) to cotton in the provinces of Bougouriba, Ioba and Poni from 2003 to 2017.

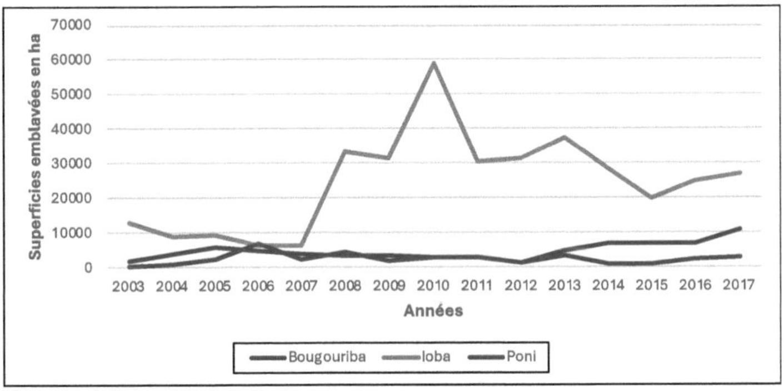

Source: Direction Générale des Statistiques Agricoles, 2019

The immediate consequence of the increase in the area sown to cotton is an increase in annual production (figure 30). Analysis of the three curves showing the evolution of seed cotton production in tonnes over the period 2003 to 2017, shows that Ioba province ranks first out of the three, with production varying according to area.

In view of the growth in production, SOFITEX granted the region a cotton ginning plant in 2000. The plant, located in Diébougou, not only facilitates the marketing of the product, but also ensures rapid distribution of agricultural inputs and seeds to farmers.

Figure30 : Evolution of cotton production (in tons) in the provinces of Bougouriba, Ioba and Poni from 2003 to 2017

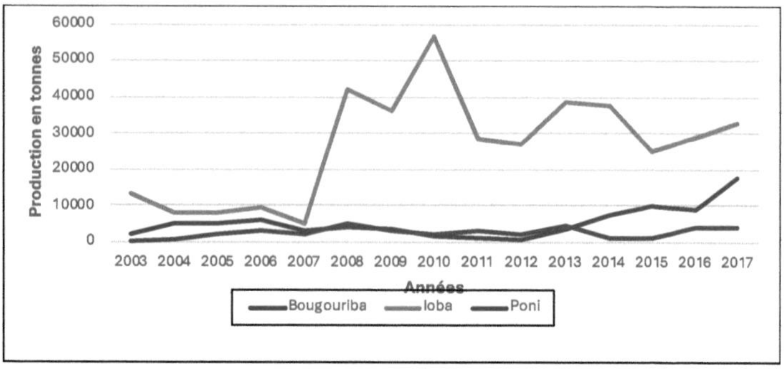

Source: Direction Générale des Statistiques Agricoles, 2019

The cultivation of cotton not only generates substantial income for cotton growers through the marketing of seed cotton, but also brings related benefits in the sense that the inputs acquired within the framework of the cotton group are used for the production of other crops, in this case maize. As part of the crop rotation practiced by some farmers, the chemical fertilizer (NPK and urea) used in a cotton field in a given year could enhance the fertility of the same field the following year for another crop with the same nutritional requirements.

On the other hand, cotton monoculture makes 'white gold' farmers vulnerable during crop years of low production due to the vagaries of the climate, attacks by cotton insect pests or falling cotton prices on the world market.

In any case, the major challenge lies in the conservation of the region's protected areas, in this case the Bontioli area, which is undergoing significant anthropization, since the conversion of forests to farmland is one of the causes of climate deterioration. In addition, the development of agriculture has led to significant changes in the state of resources such as water, air, soil, landscapes and flora and fauna habitats (Papy & Torre, 2003).

Ploughing cotton fields with tractors requires significant felling of shrubs and even trees, whereas plant communities play a role in sequestering atmospheric carbon. Pesticides, which are essential for cotton growing, are said to be the cause of pollinator insect mortality and surface water pollution in cotton-growing areas (Sawadogo, 2006; Ouédraogo, 2011).

The uncontrolled use of these inputs in cotton production has harmful short- and long-term effects on the soil. According to Dipama (2009), "the use of mineral fertilizers (NPK) and pesticides has an acidifying effect by reducing their organic carbon content. This leads to the drying out of soils through the loss of their agronomic properties, thus impoverishing them". Biodiversity could also be affected. A study carried out in the eastern region of Burkina Faso shows that cotton growers link the risks of pesticide contamination of water and fish with the risks of poisoning humans who consume them (Gomgnimbou & *al.*, 2010). The availability of chemical fertilizers is also a way for farmers to extend the areas under cultivation in plant formations. In this respect, Kaboré (2010) argues that the acquisition of chemical inputs contributes to an increase in both the areas cultivated for cotton and those sown for cereals. So how can we reconcile the challenges of growing, market-driven agricultural production with the imperative of conserving biological diversity in all its forms? Photo 4 shows a cotton field in the total reserve. In addition, the clearing of cotton fields in wooded formations (gallery forest, wooded savannah, tree or shrub savannahs) is gradually leading to their transformation into an agroforestry landscape, creating a monospecific environment in these areas.

Photo4 : Cotton field in the heart of RTFB

Field shot (SOME T. Noël, March 25, 2020)

This problem seems to oppose two phenomena that are vital to the very survival of the populations living alongside these protected areas. The solution may lie in innovation in agricultural and technological research. The use of improved cotton varieties, as is already the case for other crops, and the adoption of innovative agricultural techniques to reduce the area under cultivation and regenerate biodiversity. Organic production therefore seems to be one of the best avenues to explore, as stipulated by Gansaonré & *al.* (2020). For them, this type of agricultural production promotes environmentally-friendly practices.

6.1.3.3. Market gardening and the problem of lowland development

Market garden crops are generally cash crops grown all year round. In the dry season, they are grown around water sources (rivers, ponds, wells, marigots). The growers set up small gardens surrounded by sorghum stalks, and wood or wire fencing for the better-off to protect the plants from roaming animals. The main limiting factor for horticulture in the riverside villages is the lack of low-lying areas suitable for this type of production. This problem was regularly raised by the people we interviewed, especially in the villages of Zambo and Manoa "(...) the absence of dams and, by extension, of improved lowlands in our village is the real problem ...young men and women are obliged to migrate temporarily to neighbouring Ghana or to big cities such as Ouagadougou and Bobo-Dioulasso; those who remain are often forced to engage in gold panning or other illicit activities in the reserves (...)" said a respondent from the village of Zambo. Speculative crops such as tomatoes, cabbages, African eggplants, European eggplants, spinach, okra, sorrel, lettuce, onions, cucumbers, peppers and chillies are sold at local markets held every six days in Dagara villages, every three days in Lobi villages and every Sunday in the semi-urban centers of Diébougou, Dano, Dissin and Gaoua.

The area planted remains very small, but the negative impact seems significant in that gardens in riverbeds such as the Mouhoun and Bougouriba cause silting. Ploughing makes soils vulnerable to water erosion. The run-off of rainwater carries the loose surface part of the soil into the minor river bed, gradually silting it up. The words of an Eaux et Forêts agent stationed in Zambo clearly illustrate this analysis in the following terms: "(...) the Mouhoun and Bougouriba rivers are very silted up due to the market gardening activities on the banks of these rivers. Some gardeners even occupy part of the cattle track that runs alongside the Bougouriba river (...)". Chemical inputs (fertilizers, pesticides, herbicides) used in production are a potential source of water pollution. The gallery forests along these waterways are in places degraded by market gardening.

Moreover, article 233 of Burkina Faso's Forestry Code implicitly prohibits cultivation on the banks of watercourses: "The banks of watercourses, lakes and ponds must be protected to ensure their perimeter by the delimitation of a servitude strip on each bank or all the way around, as the case may be". Like cotton, market gardening was introduced by the colonists, and has been timidly integrated into the agrarian systems of the South-West region.

The development of lowlands in villages bordering the reserves could keep young people busy during the dry season, and act as a bulwark against the silting up of the main local watercourses. It should be noted that market-garden production on the banks of the Mouhoun River exists on the Ghanaian side, so without a synergy of action between the two countries that share the river, the actions undertaken by the Burkina Faso side would be in vain. And yet, watercourses are still home to a significant biodiversity of aquatic, forest and savannah ecosystems. Photo 5 shows market gardening on the banks of the Mouhoun River.

Photo5 : Market gardening on the banks of the Mouhoun River, left bank

Field shot (SOME T. Noël, March 22, 2020)

6.1.5. The impact of farming systems on reserve landscapes

Agricultural systems have shaped the landscapes of the Bontioli wildlife reserves. The presence of villages in the reserves has inevitably transformed these areas into agro-forestry parks made up of vast fields dotted with fruit trees, and isolated dwellings linked by tracks. The dwellings are typical of those found in the south-west. It reveals the non-hierarchical social organization of these peoples in their constant quest for independence. These trees are dominated *by Vitellaria paradoxa, Parkia biglobosa, Adansonia digitata, Lannea microcarpa, Faidherbia albida, etc.*

In the areas where the reserves come into contact with the surrounding villages, the vegetation formations are encroached upon by fields and logging operations, creating haloes of clearings at the edges of the reserves. These landscapes are gateways through which people gain quick and easy access to the reserves to carry out illicit activities such as grazing in the total reserve and gold panning. Shifting cultivation, which is a particular form of use of these protected areas, can lead to their degradation if stakeholders do not take effective protection measures.

The government of Burkina Faso has taken drastic measures to restore these two protected areas. The government's approach has been to ask people who had settled in the reserves to leave. Faced with their reluctance, the State proceeded to evict them. The ruins of old dwellings (photo 6) still bear witness to the recent human presence on these sites. Although the State has removed the populations living in certain parts of the reserves, it has not been able to prevent the quasi-permanent presence of man in these protected areas, as the same populations resettled at the end of the Projet de Gestion Durable des Ressources Forestières (PROGEREF).

Photo6 : Ruined dwelling in the total reserve

Field shot (SOME T. Noël, March 25, 2020)

The landscape still bears the marks of man. The buffer zones between these protected areas and the villages are cleared and transformed into agroforestry parks; this system of associating trees and crops is a custom in the agrarian systems of the farmers in the study area, and indeed throughout the South-West region. Trees in the fields have ecological, agronomic (*Faidherbia albida*), economic and nutritional functions. Bontoni & Reij (2009) note that trees, but also shrubs and perennial grasses, not only slow down wind speeds. They also fix sand and reduce soil temperature, thus limiting evaporation and evapotranspiration.

6.2. Pastoral production systems and their influence on plant formations

Livestock production systems involve different modes of land use, relationships between animal and crop production, and ways of adding value (Ouoba, 2013).

The pastoral system as a whole, in the communes and villages bordering the reserves, is a consumer of natural resources (herbaceous biomass, vegetation and water). There is virtually no fodder production in the locality for livestock feed. Natural vegetation is therefore the main source of grazing for domestic animals and large wild herbivores. Natural watercourses (rivers, ponds and marigots) and boreholes are the main water reservoirs for livestock watering. However, they are threatened by gold panning, an activity that uses toxic substances such as cyanide and mercury to process the minerals. Market gardening in the vicinity of these water reservoirs causes silting and degradation of the gallery forests that line the watercourses.

6.2.1. Extensive sedentary livestock farming

Traditional livestock farming is an essential component of the agrarian system in villages bordering the reserves. It is second only to agriculture as an income-generating activity.

6.2.1.1. Livestock composition

Livestock in the study area consists mainly of poultry, goats, pigs, cattle, sheep and donkeys. The number of households raising each species and the frequencies are shown in Table 32.

Table32 : Number of households by animal species raised

Livestock	Number of heads of household	Frequency (%)
Cattle	116/325	35,7
Ovine	106/325	32,6
Goat	187/325	57,5
Pig	152/325	46,8
Asinine species	2/325	0,6
Camelin	0/325	0,0
Poultry	312/325	96,0
Other	0/325	0,0

Source: Field data, March 2020

The composition of livestock is an important element to take into account when analyzing the dynamics of plant formations, since livestock farming is an extensive sedentary activity. The table shows that almost all households have livestock consisting of poultry (hens, guinea fowl, ducks, etc.), goats, sheep and cattle. With the exception of oxen, these animals are abandoned in the wild during the dry season. Crop residues and wild fodder are used respectively by 95.1% and of respondents to feed the animals, due to the virtual absence of the use of

Agro-Industrial By-Products (AIBP), of which only 0.of respondents make use.

6.2.1.2. The contribution of extensive sedentary livestock farming to the local economy

Extensive, sedentary livestock farming makes a considerable contribution to the family economy. Animals are sold on the hoof at local and communal markets. Income from the sale of livestock and poultry contributes to the family's health, schooling and other needs (food, clothing, etc.).

Cattle make an enormous contribution to agricultural production, as do draught oxen. Manure from these animals is used to fertilize the fields. Manure pits are used to decompose organic matter before spreading it on the fields.

6.2.1.3. Impact of pastoral practices on plant formations

Grazing areas are confined to the boundaries of natural areas. Of those surveyed, 44.8% stated that grazing takes place in old fallow land. Whereas 34.9% and 20.3% of respondents respectively maintained that grazing takes place on reserves and in the bush outside reserves (figure 31).

Figure31 : Grazing areas

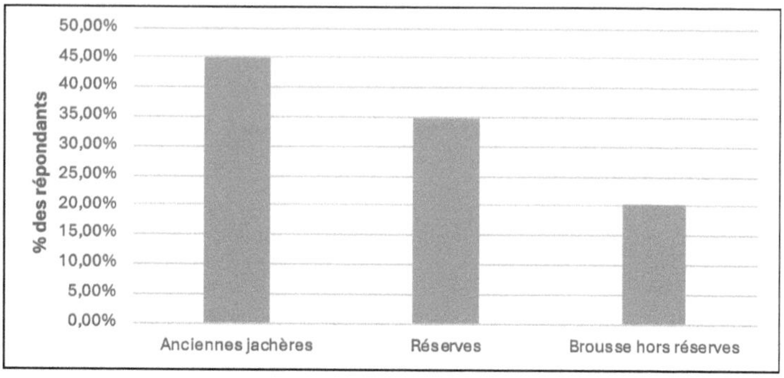

Source: Field data, March 2020

Analysis of the graph shows that livestock farming in the area is entirely dependent on natural resources, which it impacts negatively due to the relatively large presence of small domestic herbivores (goats and sheep), which have a very high destructive potential for shrubs, which they graze at every turn. Not only do they graze on buds and young shoots, but they also climb and break off small branches. In other words, they create havoc with shrubby plant formations. Cattle, although less numerous, are not insignificant in the consumption of natural resources either. A study carried out in three protected areas in western Burkina Faso by Kiéma & Fournier (2007), shows that sedentary livestock use protected areas in two short periods (April-May and mid-October to mid-November), but could at best do without them, as herd numbers remain low.

Animal straying can be detrimental to certain plant species particularly prized by animals, such as *Afzelia africana*. In the final analysis, it is animal straying that needs to be combated in order to preserve and establish the healthy use of protected areas for livestock farming. "The particularly harmful nature of the livestock-raising system based on animal roaming must be borne in mind; it makes all actions linked to tree development uncertain, as its perennial nature exposes it to the teeth of animals, especially in the dry season, when the survival of woody plants is most problematic" (Goudet, 1985). It is therefore imperative to rethink the integration between pastoral and agricultural systems, in which trees are a key element. A similar observation was made by Yonkeu & *al* (1995), in north-central Burkina Faso, specifically in the Bourzanga watershed. This study shows that in the agropastoral zone, animal rambling, largely due to the absence of cattle tracks, and certain practices such as lopping and pruning of trees by herders, have contributed to the deterioration of the woody and herbaceous stratum. In addition, 28.7% of the people surveyed in the area use tree pruning to boost animal feed.

Pastoral practices, characterized by the pruning of specific plant species to boost livestock feed, may be causing an erosion of biological diversity. Plant species such as *Khaya senegalensis, Pterocarpus erinaceus* and *Afzelia africana* are subject to intense pastoral pressure during the dry period (March and April) in the vicinity of concessions and reserves. Those questioned (farmers and transhumant herders) about the plant species pruned to feed livestock, unanimously acknowledge that they use these species. This increases the risk of a decline in the population of these plant species in these protected areas.

The strong pressure of climate change, low seed production and the slow growth of *Pterocarpus erinaceus* seedlings makes this species one of the most threatened in Burkina Faso (Dimobé, 2017). Moreover, they are listed by respondents as endangered species in reserves. The same observation of selective pruning of plant species by herders has already been reported in the Nazinon classified forest by Sawadogo (2006). He attests that prized species such as *Afzelia africana*, *Pterocarpus erinaceus* and *Balanites aegyptiaca* are regularly pruned by herders in this forest. In Bogandé and Fada N'Gourma, Karen & Thiombiano (2000) highlight the pruning of *Pterocarpus erinaceus* as a forage species. These authors claim that all individuals of this species are constantly pruned in the above-mentioned localities. The pruning of *Khaya senegalensis, Pterocarpus erinaceus* and *Afzelia africana* by sedentary and transhumant herders for livestock feed, and the harmful consequences of this pastoral practice on the sustainability of these plant species, have been reported in the Sudanian savannahs of the Atakora chain in Benin by Wala (2010).

6.2.2. Impact of transhumance on plant formations

Transhumance is not an activity of the local populations living in the Bontioli wildlife reserves. It is the work of herders from the North, Hauts-Bassins, Boucle du Mouhoun and Centre-Ouest regions, in search of pasture and water during certain critical periods of the dry season. The unavailability of pasture leads herders to fraudulently prune trees in partial and even total reserves, where all activity is prohibited. On the other hand, the transhumant herding system is dependent on the protected areas, as the herds stay there for around 5 months each year.

In the Bontioli wildlife reserves, transhumant herders spend most of their time in the dry and even wet seasons. As well as fodder for their livestock, they prune plant species such as *Pterocarpus erinaceus, Afzelia africana and Khaya senegalensis* to feed their animals. These species have little or no regeneration (Thiahoun, 2012). According to the Bougouriba province forest and wildlife manager, "transhumant herders confine their flocks between the watercourses inside the Bontioli wildlife reserves to spend the rainy season there, as they know that at a certain time of year certain areas are inaccessible. They often employ children as shepherds in the forest to prevent their animals from being impounded by forestry officials". The same practice was observed in the Koulbi classified forest. In this forest, transhumance takes place clandestinely, but is easily observed through the cutting and mutilation of woody fodder (Tankoano, 2012).

Transhumance periodically increases the stocking of livestock at a time when vegetation growth is at its lowest, in the dry season. The intensive use of reserve plant resources by stockbreeders therefore leads to the regression of certain plant formations. Grazing, combined with regular trampling of trees and especially shrubs by the animals, leads to the evolution of certain shrub formations towards totally degraded or bare areas. Livestock grazing therefore appears to be a major factor in the degradation of plant formations in these protected areas. Photo 7 shows a herd of oxen grazing in a cotton field in the total reserve.

Photo7 : Transhumant herd of cattle in a cotton field in RTFB.

Field shot (SOME T. Noël, December 28, 2020)

The outlook for livestock farming, in a context of resource scarcity where the trend is towards conservation, is that modernization is a necessity. Intensifying pastoral production, by producing fodder for livestock and using modern techniques such as cattle fattening, would reduce the anarchic consumption of vegetation by roaming animals. Breeders could thus create a more intensive agropastoral system in terms of farming, combining agricultural, pastoral and forestry production.

6.3. Farmers' perceptions of changes in reserve plant formations

Quantitative assessment of plant resources by farmers is not an option. On the other hand, the aim here is to understand their overall assessment of reserve dynamics and the reasons behind it.

6.3.1. Reserves: State expropriation of riparian populations' land?

The State's classified forest estate was created by the French colonial administration without any real knowledge of the social structures established in these localities. The land on which the reserves were located was considered, a priori, to be wild and without masters. The areas selected for their hunting, climatic and economic interests were considered by the colonizers as vacant land, even though they belonged to families. Land was never masterless. It always belonged to spirits to whom the first occupants erected an altar before settling and exploiting it (Hien, 2000).

In fact, most of the villages now engulfed by the partial reserve were farming hamlets established on their current sites before the reserve was created. As the population grew, these farming hamlets became administrative villages. Socio economic infrastructures (dispensaries, markets) and educational facilities (schools) were set up for the benefit of these populations.

Today, the demographic growth that has led to these farming hamlets becoming administrative villages, and the change in the agricultural production system - still extensive but more space-consuming - have given rise to perceptible competition for land. As a result, indigenous populations are claiming ownership rights to these forest lands. In addition, article 6 of the decree classifying the partial reserve specifies that the communities of Zambo and Bapla within the partial wildlife reserve are recognized as having customary rights to clear, cultivate, cut wood for personal use, herd grazing, fish and gather fruit and plants for food, medicine and dye, subject to compliance with current regulations (MECV, 2009).

Moreover, customary land law applies to land management. The land chief is obliged to find a plot of land for a foreigner who requests it.

The status of the partial reserve is currently being contested by the local population. In this respect, the 1996 eviction by the Burkinabe state from the reserve was perceived by the communities concerned as a violation of their rights. "(...) they came to destroy our homes, burn our granaries, chase us away, it's not normal (...)" confided one respondent. The chief of the village of Manoa added: "This situation has created temporary and even permanent migrations towards neighboring Ghana, where the King of Lora has welcomed and settled a large number of these migrants".

A feeling of expropriation of land by the State is therefore emerging in peasant opinion. In our interviews with farmers at , some even asked the State to declassify part of the partial reserve, if only their current fields, to enable them to produce. "(...) the reserve is unused and therefore worthless, and we're looking for land to feed our children", a Bobra-Gogo farmer told us. So what use is a resource to a vulnerable population if it can't provide them with their basic needs? The main concern for the population is not the ecosystem functions that the reserve provides, but what they can get out of it for their very existence. This observation is similar to that of Gomgnimbou (1998) in the classified forests of Tiogo and Laba in central-western Burkina Faso, where farmers are demanding that the state declassify the forest for agricultural production. Fournier & Millogo-Rosolodimby (2007) report on the sense of injustice felt by local populations in situations of administrative restrictions on the use of protected areas in western Burkina Faso.

6.3.2. Farmers' knowledge of the evolution of plant formations in reserves

According to 68. of those surveyed, the reserves are undergoing a regressive evolution. On the other hand, 11.7% claim that they have evolved positively; 11.had no perception of the evolution of the reserves' plant formations. Finally, for 8% of respondents, there has been stagnation. These data show that the population has a good knowledge of the evolution of plant formations in the reserves, since they are responsible for most of the human activities carried out in these protected areas.

6.3.3. Reasons for the decline in reserve plant formations

According to the respondents, the main reasons for the regressive evolution of plant formations are anthropogenic and natural (see figure 32). In terms of the origins of this regression, 26. cited the cutting of green wood, 21.the opening up of fields for agriculture, 18.late fires, 5.gold panning and 2.grazing. As for natural causes, 14.5% of those questioned said that drought was responsible for the decline in plant formations; 4.for strong winds uprooting large trees, and 1.for high temperatures.

Figure32 : Reasons for the decline in reserve plant formations

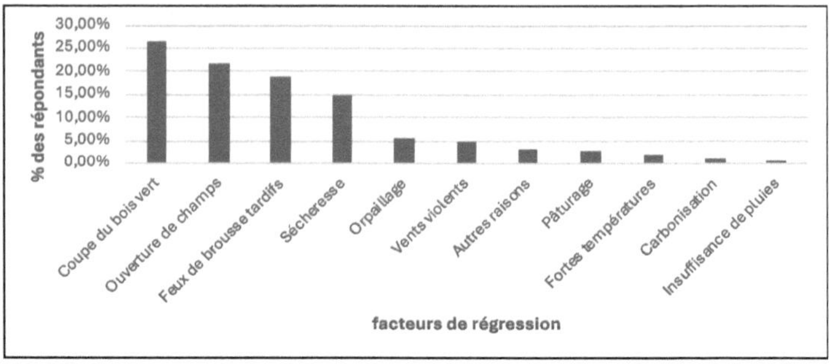

Source: Field data, March 2020

Analysis of the graph shows that the main causes of the degradation of plant formations listed by the local people are human. This result corroborates the observation made in the partial reserve by Somé (2013) and in the total reserve by Dimobé (2017). Several studies carried out in Sahelian countries confirm this observation (Ozer & Ariori, 2005; Ozer, 2006; Botoni & Reij, 2009).

The fast-growing populations of the study area have implemented agrarian systems that have undergone various mutations. Farming systems remain generally traditional, due to the use of techniques that do not favor water and soil conservation, and rudimentary tools employed by farmers. Agricultural productivity is therefore low. However, cotton production is more demanding and requires the use of relatively modern methods and tools, whose products are exclusively market-oriented, and has a higher productivity than cereal growing. Extensive farming in the context of saturated agricultural land is damaging the reserve's plant formations, which are continually being encroached upon by fields. Pastoral practices linked to the selective pruning of plant species, grazing in protected areas, the almost exclusive use of wild fodder, and the rambling of animals from extensive sedentary and transhumant herding, characteristic of the zone's pastoral system, have a negative influence on the evolution of woodland formations and plant biodiversity in the reserves.

CHAPTER 7: RESERVE MANAGEMENT AND PROTECTION

The emergence of the concept of sustainable forest management raises the question of how forests and woodlands are managed around the world. Forest management is the process of planning and implementing practices for the responsible management and use of forests and woodlands to achieve specific environmental, economic, social and cultural objectives (FAO, 2005)[18] . Traditional methods of exploiting forest resources are increasingly being criticized, and are therefore gradually being abandoned by some rural communities, whether willingly or not. The aim is to analyze the management methods applied in the Bontioli wildlife reserves, and to identify the dysfunctions and protection strategies of these reserves.

7.1. Reserve management methods

In principle, management of the Bontioli wildlife reserves should respect the status of each. In view of the statutes conferred on them by their creation decrees, which have not officially changed, it would be incongruous to speak of management stricto sensu of these protected areas in such a context. Indeed, the notion of management refers to the establishment of a structured system (traditional or modern), recognized by the members and aimed at a common objective, which in this case is the sustainable exploitation of forest resources. In practice, the main users of these protected areas do not seem to have fully assimilated the philosophy behind their creation.

[18] https://www.fao.org consulted on 10/11/2022

7.1.1. Changes in reserve management

Created to ensure total and partial protection of wildlife, the management of these two forest entities has evolved in fits and starts according to political peripateties through the intervention of state projects and programs. Several periods have punctured this management.

- **From 1957 to 2004**

For the conservation of biodiversity, protected areas are privileged tools for protecting wildlife (Binot & Joiris, 2007). It should be noted that there has never been any formal management involving the local populations living in these protected areas since their creation in 1957. In their quest for survival, these populations have always engaged in agro-pastoral activities, gathering, hunting and fishing in the reserves. For them, forest products provide a secure livelihood. These forest enclaves still abound in arable and grazing land, so producers are easily attracted by the availability of these resources.

The management system put in place by the State was dirigiste and mainly conservation-oriented. State intervention was therefore limited to surveillance, protection, reforestation and anti-poaching measures. These actions carried out by the Water and Forestry Department failed to achieve the desired objectives. In 1996, they resulted in the forced eviction of people illegally occupying the reserves. To solve the problem of strong pressure from local populations on these protected areas, the Bontioli Partial Wildlife Reserve was granted a concession on March 10, 1997 as a game ranch along the lines of the Nazinga ranch, with a view to promoting wildlife in the South-West. To manage the ranch, twelve (12) Comités Villageois de la Gestion de Faune (CVGF) were set up in twelve riverside villages. However, the ranch did not operate due to the death of the promoter (MECVCC, 2017).

- **From 2004 to 2012**

No natural resource can be managed sustainably unless the communities that use it are well organized. Whereas in most protected areas in Burkina Faso, conservation depends on time-limited projects and external donors (Kiéma, 2007). Nonetheless, they do help to strengthen the organizational capacities of forest resource management stakeholders and to finance certain management, protection and forest product exploitation activities. It was against this backdrop that the Projet de Gestion Durable des Ressources Forestières (PROGEREF) in the South-West, Centre-East and East Regions was launched in these protected areas in 2004, with the specific aims of: (i) improving the management of forest and wildlife resources; (ii) increasing local incomes. It is structured around three components:

- ✓ component 1: sustainable management of forest resources ;
- ✓ component 2: capacity building and accompanying measures ;
- ✓ component 3: project management.

The first component is subdivided into two sub-sectors: forest and wildlife. The specific objectives assigned to the forestry sub-sector can be summarized as follows: (i) to reduce anthropic pressure on forest resources and curb their degradation; (ii) to ensure the reconstitution of biological diversity; (iii) to ensure sustained productivity of forest products; (iv) to generate jobs and contribute to increasing household incomes. With regard to the wildlife sub-sector, the specific objectives assigned to it are: (i) to promote and energize village wildlife management organizations, taking into account traditional modes of organization; (ii) to promote sectors such as ranching and small-scale wildlife farming, and the reintroduction of local species; (iii) to strengthen the protection of hunting resources; (iv) to develop vision tourism and sport hunting through hunting zone concessions (MECV, 2009).

The second component aims to :

- contribute to improving agricultural production ;
- improve access to basic social services and social protection ;
- expand employment opportunities and income-generating activities for local populations;
- strengthening local governance ;
- contribute to better management of transhumance ;
- strengthen technical and technological capabilities.

Notwithstanding the ambitious objectives set by PROGEREF, its actions have covered only a few areas. They concern:

- support for RPFB monitoring;
- joint diagnostics in 2006, involving seventeen (17) villages bordering the RPFB ;
- large wildlife inventory;
- floristic inventory ;
- reclaiming degraded land by scarifying and reforesting 278 ha.

The preliminary project for the development of the Bontioli wildlife reserves drawn up by PROGEREF and due to be carried out in two phases, the first of which ran from 2010 to 2014, has never been validated by state actors (MEEVCC, 2017).

- **From 2012 to 2017**

The 2012-2017 period was marked by an absence of project/program intervention in the reserves. Flagship protection activities were limited to those of the forestry service.

- **From 2017 to 2021**

The Forest Investment Program (PIF) intervened in the Bontioli wildlife reserves with the following main challenges:

- ensure the conservation of forest and wildlife resources in and around reserves by securing land resources;
- contribute to improving biological diversity and combating desertification ;
- increase the contribution of forest and wildlife resources to the local, regional and national economy and to the fight against poverty through the development of activities to add value to the forest and wildlife resources of the two reserves;
- ensure good governance in the context of decentralization with a view to improving transparency in the management of forest and wildlife resources (MEEVCC, 2017)

In the course of its work in these reserves, the program has instilled a new dynamic in the organization of villagers into management structures. Two Forest Management Committees (CGF) have been set up for the two reserves. These committees comprise ten (10) members from each commune. The RTFB CGF is made up of thirty (30) members from the communes of Zambo, Dissin and Diébougou. On the other hand, the RPFB CGF has forty (40) members from four (4) communes, namely Nako, Bouroum-Bouroum and Tiankoura. These two structures have each set up a twelve (12) member bureau to coordinate operations, assisted by two (02) advisors.

Alongside the forest management committees, other structures exist, such as associations of hunters and NTFP processors.

On the whole, the level of organization of local populations for the efficient use of forest products from these protected areas remains very low, despite the existence of the beginnings of farmers' organizations for their participative management. The structures put in place are not representative of the diversity of users of the resources of these protected areas.

Key activities carried out in the reserves, such as green wood harvesting, agriculture and livestock breeding, which should motivate and guide managers in the integrated management of resources by all stakeholders, are not really taken into account in this organization. With this in mind, a significant proportion of those surveyed (93.) admit that they are not involved in reserve management. These people cite reasons linked to the selection of members (64.6%), poor collaboration with foresters (6.), disinterest (), and other reasons (2.). A very large number of people are therefore not involved in forestry activities because of the selection of members who are forest supervisors and focal points. This selection was made during the implementation of the Forest Investment Program (PIF). The non-management of reserves is seen as a major constraint to the participation of local populations in the sound management of their resources.

The activities carried out by these structures, with the exception of hunting and NTFP processing associations, do not generate income for local populations. They are essentially oriented towards protection and restoration. So, to encourage the abandonment of certain practices such as bushfires, the PIF has supported these management committees with an annual remuneration of 3,000,000 FCFA if the delimited area has not been burned until the next winter season; 2,000,000 FCFA if a third of the area has been burned and 1,000,000 FCFA if two-thirds has been burned. He has also chosen two people, notably young people from each village, to act as "forest supervisors". They are equipped with means of communication (cell phones) and bicycles. The mission of the "forest supervisors" is to support the forestry service in protecting the reserves. The main activities organized by these actors are awareness-raising and reserve surveillance, recognized by 67.and 46.5% of respondents respectively. Other activities are somewhat marginal. The opening of layons, early burning, reforestation and other activities initiated by them are recognized by 4., 7., 7.1% and 4.6% of respondents respectively.

The financial resources generated by the issuance of operating permits to fishermen, deadwood collectors using carts and tricycles, and hunters during the hunting season, are paid into the public treasury. In addition, two categories of hunting permits are issued: a permit for small-scale hunting, which costs 15,000 CFA francs for nationals and 35,000 CFA francs for resident expatriates; and 70,000 CFA francs for non-resident expatriates. For big game hunting, the license is set at 30,000 F CFA, 80,000 F CFA and 620,000 F CFA respectively for nationals, resident expatriates and non-resident expatriates. The mixed hunting license for nationals is 35,000 F CFA.

In parallel with the PIF's activities, the forestry department has pursued its regalian activities of protecting and restoring reserves, and issuing permits for the exploitation of forest products in accordance with current regulations. Thus, 58.10% of respondents (population) felt that these agents organized forest surveillance activities, while 38.50% responded in favor of reforestation, compared with 3.for other activities, i.e. awareness-raising (figure 33).

Figure33 : Protection and restoration of reserves by state structures

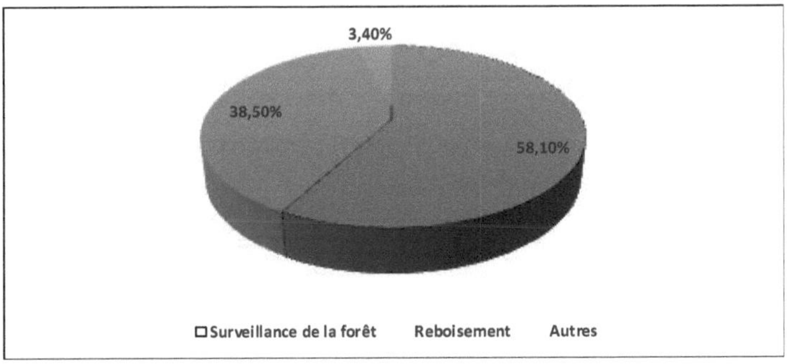

Source: Field data, March 2020

An analysis of the graph shows that the State, through its deconcentrated departments, places the emphasis on protecting reserves from anthropic pressures.

During these periods, the exploitation of forest resources was virtually permanent. Indeed, the cutting of green wood, agriculture and transhumance in the reserves have always been permanent practices that have not been organized in any significant way, either by the State or by the actors themselves.

7.1.2. Harvesting forest products

With the exception of timber and wildlife, which are harvested on a permanent basis, other forest products are harvested on a seasonal basis, in line with the phenology of the various plant species present in the reserves.

7.1.2.1. Wood harvesting

Wood is essential to the lives of the communities living in the study area. Almost all households use it for culinary and commercial activities, in this case the manufacture of local beer commonly known as "dolo". Its use is almost permanent.

Green wood is generally cut in reserves, fallow land and fields. However, a local practice known as "daartchè" in the local Dagara language, which consists of building up security stocks of firewood, is widespread in the locality. This practice is carried out by women at the end of the rainy season, when they are relieved of their work in the fields, and shrub and tree shoots are not damaged by bush fires. They use machetes to cut down shrubs and limb certain trees. The practice of cutting wood, particularly shrubs, by women has been reported in the localities of Sampéto in Benin by Bonou (2008) and in Barh-Kôh in southern Chad by Madjimbe & *al.* (2018).

The collection of firewood has a negative impact on the reconstitution of plant cover and biodiversity, as large trees are lopped and young ones cut with no respect for the cutting technique designed to encourage their regeneration. Species such as *Combretum glutinosum, Combretum nigricans, Combretum molle, Combretum fragrans, Combretum collinum, Crossopteryx febrifuga, Diospyros mespiliformis, Anogeissus leiocarpus, Lannea microcarpa*, etc. are the most commonly exploited by women. Photo 8 shows plants cut in the shrub savannah of the total reserve.

Photo8 : Trees cut down at RTFB

Field shot (SOME T. Noël, December 24, 2020)

The cut wood is stored in safety stocks near the concessions or on the outskirts of the villages (see photo 9), to be used generally during the rainy season, when dry wood is in short supply. This means that women, who are very busy with field work at the time, do not have enough time to search for wood in the bush.

Photo9 : Stock of wood on the outskirts of Zambo

Field shot (SOME T. Noël, December 24, 2020

Dead wood is also taken from reserves by men and women for a variety of uses. These include trees that have died as a result of attack (termites) or been burnt by fire, uprooted and dried out, or dry branches.

7.1.2.2. NTFP collection

NTFPs are frequently collected individually or in groups by the various women's groups and the Union des Coopératives de transformation des PFNL based in Djikologo. It takes place on a permanent basis in the reserves, depending on the ripening period of the fruits of each plant species, and temporarily for honey, which is collected in the dry season. There are no restrictions on the harvesting of these products in the reserves. The same observation was made by Ibrah (2005) in the department of Filingué in Niger concerning the collection of jujube (fruit of *Zizyphus mauritiana*), and gum arabic (fruit of *Acacia senegal* and *Acacia seyal*).

Various materials and techniques are used in NTFP harvesting. These include the lopping of thorny trees such as *Balanites aegyptiaca* for the leaves, and the use of poles to pick fruit perched on branches far from the ground. This is the case for *Parkia biglobosa* and *Adansonia digitata. Vitellaria paradoxa* fruits and nuts are usually picked by women and children. They climb trees to collect the products they need (fruit, flowers or leaves). This is a risky business, as accidents often occur, resulting in serious injury or even death. To collect bark and tubers (roots), the equipment used is an axe, a cutter and a hoe for skinning and desoiling.

NTFP harvesting methods based on gathering/gauling, delimbing and debarking have been reported in Arly Park in the eastern region of Burkina Faso by Doamba (2012). The various techniques used to harvest NTFPs have harmful consequences for the survival of certain species, which are overexploited because of their utilitarian nature. However, measures exist to curb the phenomenon, even if they are proving ineffective in the field. According to the head of department at the Djikologo forestry post, these include "raising public awareness of the need to protect utilitarian species and reserves, and patrols organized by foresters to crack down on prohibited practices".

7.1.2.3. Honey production

Honey is traditionally produced on the reserves from local beehives. These are canaries or pots made from baked clay and stored in tufts. These hives are pre-coated with water from the processing of *Parkia biglobosa* (néré) seeds to rapidly attract the bees. In both reserves, the presence of bees is low (personal observation in the field), probably due to the use of pesticides in cotton production and chemicals (mercury and cyanide) used in gold processing. Honey is collected twice a year on average, with production quantities proportional to the volume of the hive. In the absence of modern equipment, fire is the safest way to collect honey. It does, however, have adverse effects on the quality of the finished product, since it destroys certain nutrients. The use of fire to collect honey also causes forest fires.

7.1.2.4. Exploitation of wildlife products

Despite the progressive loss of habitat, the reserves are still teeming with important wildlife resources. According to IUCN/PACO (2009): Elephant, Hippotrague, Guib Harnaché, Cobe de buffon, Ourébi, Redunca, Bubale, Warthog, Baboon, Orycterope, hare, etc.

Hunting activities take place regularly in protected areas. There are two types of hunting in the area: traditional hunting and hunting authorized under the terms of the classification. This consists of a battue during which hunters use bladed weapons (bows, clubs, knives, slingshots, etc.) and fire to hunt small game (hare, francolin, partridge, bucks, etc.). Battues are organized at the height of the dry season, after the harvest, when the herbaceous biomass, which is very abundant in both reserves, is completely dry, and the fires lit in these circumstances cause large-scale fires. A study of bushfires in western Burkina Faso by Valéa (2010) confirms that protected areas are the main places where large areas are burned. This is due to the continuity and availability of herbaceous biomass as a fuel. Traditional hunting fires destroy wildlife habitat.

Professional hunting using firearms is controlled by the Ministry of the Environment, which issues hunting permits to hunters. Hunters' organizations can be found in certain communes bordering the RPFB. There are three in particular: the Association Dougougnéko des chasseurs dozos de Wangara in the rural commune of Tiankoura, the Comité de Gestion de la Chasse Laonmarlé in the rural commune of Zambo, and the Comité de Chasseurs Assiyko Di Ha in the rural commune of Nako. These peasant structures, equipped with hunting permits, carry out their activities during the legal period corresponding to the national hunting season. Article 123 of Burkina Faso's 2011 Forestry Code stipulates that no one may carry out an act of hunting unless they hold a hunting permit or a special authorization issued by the minister responsible for wildlife. These structures contribute to the protection of plant and wildlife resources. The animals slaughtered include the hippotrague, the zebra, the buche, the guib harnaché, the hartebeest, the warthog, the hare and others. This regulatory provision is not always respected by many players in this sector. Due to the proximity of the reserves to the Republic of Ghana, which forms a natural boundary (Mouhoun River), control remains delicate.

7.1.2.5. Harvesting fish products

Fishing activities are organized around the main rivers (Mouhoun and Bougouriba) for around three months of the year by professional fishermen known as "bozos", some of whom have come from the Republic of Mali, and traditional fishermen. Net fishing is the most common method used by professionals, while others use creel fishing, hook fishing, beating the water, etc.

A professional fishing camp exists in the village of Manoa. Entire families are mobilized here, with the men in charge of casting the net and bringing in various species of fish. The main species caught include *Tilapia zillii* (carp*), Claria niloticus, Lates niloticus* (skipper)*, Clarias* *gariepinus* (catfish), *Oreochromis fasciatus, Heterotis sp, Protopteris anectens*, etc. The women work hard on processing tasks, which involve smoking fresh fish in traditional ovens. This technology requires the use of large quantities of firewood. Drying fresh fish is a processing technique little used by local fishermen. Fish products are therefore sold fresh or processed in local markets and in certain localities in the Republic of Ghana, close to the production zones.

Numerous harmful practices are hampering the emergence of fishing in the locality. These include the silting up of watercourses due to the degradation of banks by agricultural and pastoral production, and water erosion. Water reservoirs dry up in the dry season. Gold panning poses a real threat to aquaculture resources in the area, due to the use of prohibited toxic products (cyanide, mercury, zinc) in ore processing, which contribute to the pollution of water points.

Some fishermen use mosquito nets, even though these practices are prohibited by article 175, paragraph 4 of Burkina Faso's current Forestry Code: it is forbidden to "use fishing gear made of mosquito cloth or netting to capture fry". The hawksbill net is the method best suited to Burkina Faso's aquatic ecosystems (Ouédraogo & *al.,* 2015). It is a conical piece of gear with weights at the base that the fisherman throws so as to deploy a pouch that caps the fish (FAO, 2008).

7.1.3. Analysis of reserve management methods

The evolution of management methods in the Bontioli wildlife reserves is intimately linked to the overall conception of protected area conservation in Burkina Faso, and also to that of international paradigms of forest resource management. The state's monopoly on management has created competition for resource control with indigenous populations. As a result, the conflict arising from the divergent interests and perceptions of the two actors (state and population) has a negative impact on plant formations.

With the intervention of projects and programs, especially the PIF, in these protected areas, the involvement of the population in the management of their reserves has gradually begun to become a reality. At this stage, however, local participation remains embryonic. Indeed, their main interests are still not taken into account, as there is no integrated and coherent development program for agriculture and livestock farming for the benefit of these mainly farming populations.

The structures set up to support the forestry administration do not have a budget for their operations, as there is no key for distributing the funds generated by the issuing of logging permits between the State and them. With their activities almost entirely devoted to protection, these management committees carry out voluntary work. Innovative financing mechanisms should therefore be explored. In addition, the institutional framework does not provide sufficient opportunities for these committees and the local population to manage the products of the reserves in a sustainable way, as they are often exploited clandestinely by groups with widely divergent interests. The forest mobilizes a multitude of users, and each group of actors has its own system of production and social representation of the forest, which varies according to the issues at stake (Akouehou, 2004).

7.1.4. Implications of management methods on reserves

Reserves are subject to a variety of anthropogenic pressures. For 74. of respondents, the irregular activities carried out there are agriculture, followed by green wood cutting (23.), bush fires (13.), livestock farming (8.) and gold panning (6.). Agriculture remains the main activity practised in these protected areas. According to the people surveyed, 97.5% of them devote themselves to this task as their main activity. Other activities, such as poaching, are practised insidiously.

7.1.4.1. Poaching and the threat of extinction of large game reserves

Poaching is insidiously practised in the Bontioli wildlife reserves, where it takes two forms: local or national poaching and cross-border poaching. The former is carried out by Burkinabè outside legal hunting seasons and/or on animals protected by forestry regulations. Corruption among local political leaders and administrative officials is hampering efforts to combat the practice. "(...) a convicted poacher was released a few days after his incarceration in Diébougou under the aegis of a political leader," asserted a forester during an interview. This statement is corroborated by that of IUCN/PACO (2009), which denounced the interventionism of politicians in favor of offenders. In this regard, she affirms that a letter has been sent to the Prime Minister by the local population concerning their previous eviction.

The second is carried out by hunters from the Republic of Ghana. An Eaux et Forêt agent stationed in Zambo explains: "(...) they come to set up a space in the reserves and hunt large animals for at least a week, during which time they dry the meat and return, but we have not yet apprehended them (...)". Insufficient material and financial resources, low forestry staffing levels and terrorism make the fight against poaching in these reserves even more difficult.

In fact, pachyderms have almost disappeared from the reserves due to agricultural colonization and poaching. Elephants are regularly cited by respondents as an endangered animal species in the reserves. Rare or threatened species have been identified by IUCN/PACO (2009) in these protected areas. These are : Bubale, Cobe Defassa, Guib Harnaché and Redunca. Poaching undermines wildlife protection measures and consequently all the benefits derived from them.

7.1.4.2. Gold panning: a dangerous activity for the reserve environment

In their quest for economic and social well-being, local populations engage in a wide range of activities, often at great risk to the environment and their own existence. Traditional gold mining is expanding rapidly, to the detriment of the South-West region's natural resources.

In the Bontioli wildlife reserves, gold panning poses a dangerous threat to soil, plant and water resources. It is practised for the sole benefit of the operator, but to the detriment of "everything". The consequences are immeasurable in terms of mineral extraction and processing. There are three categories of gold digger," confided one. "There are those who have machines to search for gold in reserves, those who dig holes ranging from a few centimetres to around a metre or more, and those who dig several metres vertically, then horizontally in search of gold veins".

The process of processing ore into marketable gold is a long and arduous one. Once the stones containing the gold seeds have been extracted, they are crushed and ground in mills or crushed in iron mortars, reducing them to a powder. This powder is then washed with water to separate the gold grains from the sand. The gold residues thus obtained are treated with mercury or cyanide to obtain marketable gold. Mercury is fraudulently imported from towns in Burkina Faso and neighbouring countries.

These artisanal gold miners seem aware of the environmental risks of using products to process gold. This is why they dig large holes at some distance from the stream to wash the gold. All in all, as the washing activity regularly takes place in the major bed of the watercourse, heavy metals could be released by the processing of this mercury-, zinc- and cyanide-based ore, contaminating groundwater and surface water through infiltration, percolation and runoff during the rainy season. These processes are linked to climatic conditions, topography and soil type, which determine the hydrodynamic regime and therefore have a considerable influence on the movement of water and its content of dissolved and suspended substances (Calvet & *al.*, 2005). Excessive use of these products can lead to chemical pollution and subsequent destruction of the soil structure. Some gold comptoirs are relocated to the outskirts of villages with the same treatment process. In this case, water from pumps and wells is used. The risk here is the probable contamination of groundwater, which could poison wells for human or animal use.

The environmental impacts of gold panning are also characterized by the disruption this activity causes to the ecosystem. Yaro (2013) speaks of the disfigurement of the landscape. The holes scattered around the sites cause huge risks of domestic animals and wildlife falling. These holes are dug in wooded areas after the plants have been exterminated from their environment and the soil stripped.

Furthermore, the noise generated by the shredding machines installed in the reserves disturbs wildlife, whose habitats are already under threat. This finding confirms that of Affessi & *al.* (2016) in the Bounkani region of Côte d'Ivoir. These authors highlighted the destruction of plant cover and the negative impact of noise on wildlife.

Trees are cut down in search of gold, because if the "gold detecting" machine signals the probable presence of this metal even between the roots of a tree, they destroy it. As some gold miners have revealed, there are elements that indicate the presence of gold, namely *Khaya senegalensis*, quartz and residues from ancient iron metallurgy. This statement is corroborated by Mama et *al.* (2020), who affirm that gold is concentrated in angular and unclassified scree, quartz, shale and sandstone. Searches are therefore very often directed towards these areas.

The gold miners are well aware of Article 96 of the Forestry Code, which prohibits gold panning in reserves. They justify their activity by the lack of financial means and especially by unemployment. They are unanimous about the poor relations they have with Eaux et Forêts agents: "They take away our working equipment and motorcycles when they apprehend us", they say.

Moreover, the gold miners have not carried out any restoration work, as they claim they never reforest areas subject to gold mining. Photo 10 shows earth dunes and gold panning holes in the partial reserve.

Photo10 : Environment disfigured by gold mining in the RPFB

Field shot (SOME T. Noël, December 28, 2020)

7.1.4.3. Vegetation fires: what is at stake for plant formations in reserves?

Vegetation fires are characteristic of savannah ecosystems. They are necessary to maintain the plant formations of these ecosystems.

7.1.4.3.1. Those involved in wildfire

The existence of vegetation fires was confirmed by 65.5% of respondents, compared with 26.who disagreed and 8% did not. These vegetation fires or common fires have been reported in several savannah formations and protected areas in Burkina Faso, where the herbaceous biomass favors their spread (Valéa, 2010; Savadogo, 2009; Fournier & *al.,* 2012).

Vegetation fires are a real phenomenon in the reserves. However, they remain a highly sensitive issue within local communities, as the perpetrator of a bushfire is punished by article 264 of the Forestry Code (2011), which states that: "Sont punis d'un emprisonnement de trois mois à deux ans et d'une amende de cinquante mille (50 000) francs CFA à un million (1 000 000) de francs CFA ou de l'une de ces deux peines seulement :

- those who light uncontrolled fires in protected forest areas;

- those who, through negligence or carelessness, cause uncontrolled fires in classified areas".

Figure 34 summarizes the perpetrators of vegetation fires in the reserves by population.

Figure34 : Persons responsible for vegetation fires on reserves

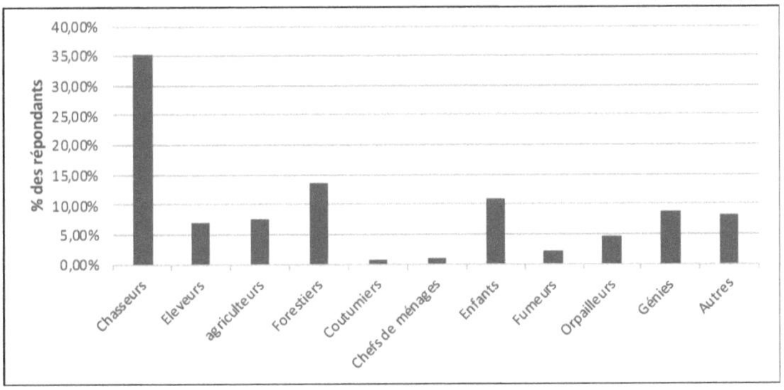

Source: Field data, March 2020

Apart from foresters who organize early fire sessions, the main perpetrators of vegetation fires are the local population. A belief in the mysterious origin of certain vegetation fires, for which genies are responsible, is widespread in the locality. "(...) it's the genies who burn the bush (...) when the time comes, at night, we see them on the hill lighting the fire, but we don't dare go near them (...)" confided one respondent. Whether it's a case of shirking responsibility or a mysterious act, the real problem with vegetation fires, especially in reserves, is the time of year they occur.

7.1.4.3.2. Vegetation fire periods

The determining factor in vegetation fires is the period of passage. Early fires occur just at the end of the rainy season. The timing of early fires depends on the length of the rainy season. Determining the soil dryness index helps identify the period of early fire.

In the context of the study area, this period corresponds to late October-December. At this time, as the herbaceous biomass is not totally dry, the spread of fire by wind is limited. As a result, the soil's surface organic matter is not affected. Traditionally, early fires can start when, depending on the area, the leaves of certain woody species such as *Lannea acida, Terminalia spp, Sterculia setigera* and grasses such as *Loudetia togoensis* begin to turn yellow (PREDAS, 2005). This is the time when fire does less damage to plant formations. After the passage of the so-called early fires, new regrowth is observed on the fire tracks. According to Sawadogo (2009), these young shoots are a major source of food for large wild herbivores. Transhumant herders also move into the reserves in search of these young grasses.

The overall trend in vegetation fire management policy in Burkina Faso is towards early and management fires. There are two reasons for this choice: "The first is utilitarian: these fires form firebreaks for more violent and devastating fires. The second is more ecological: early fires are less devastating because they act when vegetation water stress is low (at the end of the rainy season)" (Caillaut & *al.,* 2009).

Fires burning from January to May are known as "late fires". The beginning (January) is only indicative, as the dryness of the soil depends on several parameters, including the date of the end of the rainy season, and this date varies greatly from season to season. As the herbaceous biomass is completely dry, as is the soil, the spread of fire is accelerated by the harmattan. Fire can therefore ravage a large area in a very short space of time. Its effect on soil structure is that it destroys the soil's superficial organic elements. It destroys colloidal humus, which becomes mineralized, leading to structural disintegration (Guira, 1997).

Fire also destroys the microfauna, which plays a non-negligible role in soil formation and development. One of the damaging consequences of fire on plant formations is that its intensity is such that it can destroy seedlings, diseased trees, flowers and fruit. "In terms of flora, fire selects certain resistant species (pyrophiles) to the detriment of others less well equipped to defend themselves (pyrofuges). The shoots of certain trees are more or less severely damaged by fire, which modifies the habit of these trees and affects the very physiognomy of plant formations" (Valéa & Ballouche, 2012). Photo 11 shows the effect of fire on the vegetation of the total reserve.

Photo11 : Vegetation fire in the RTFB

Field shot (SOME T. Noël, December 27, 2020)

In the study area, 42.7% of respondents attest that wildfires occur on the reserves during the October-December period, while 33.believe they burn between January and February. The final group of respondents (23.) on this question maintain that they are recorded between March and early May (see figure 35).

Figure35 : Bushfire periods in the reserves

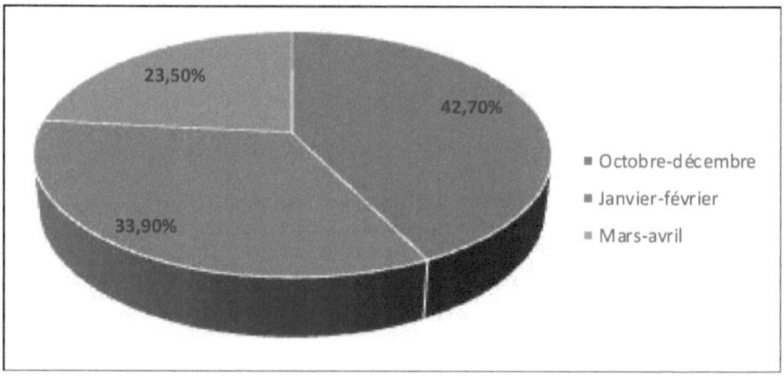

Source: Field data, March 2020

Analysis of the graph shows that reserves are affected by fire throughout the dry season. They are particularly destructive during the harmattan season. Plant formations are therefore under the influence of fire, which hinders the evolution of seedlings into trees and the maintenance of certain fragile trees that are not fire-resistant. Fires compromise not only the natural regeneration of the shrub layer, but also the regrowth of new shoots (...). They burn the herbaceous cover and certain trees eaten by livestock, thus compromising their survival during the dry season (Solly & *al.*, 2020).

7.1.4.3.3. Reasons for vegetation fires on reserves

There are a variety of reasons for the presence of vegetation fires on the reserves. Of those interviewed, 43.2% confirmed that fires are set on reserves for hunting purposes, 35.for unknown purposes, 8.to clear fields, 6.to keep reptiles away from dwellings and 6.inadvertently.

Vegetation fires periodically shape these areas, which should be protected. Only "management" fires should be allowed in these reserves, but local people burn the forest for their own reasons. On the subject of vegetation fire motives, Valéa (2010) notes that, in temporal terms, the fire calendar is linked to the agricultural and pastoral calendar. She argues that in shifting agriculture, fire is the quickest, simplest and most efficient way to clear fields. This is borne out by the reasons given for setting fires in the villages bordering the reserves. What's more, the vast majority of farmers continue to practice mining-type agriculture, in which fire is an integral part of field management.

As mentioned above, the sensitive nature of vegetation fires in the study area is very noticeable. This fact stems from the repressive behaviour of Eaux et Forêts agents and the State in general, linked to the displacement of these populations. The survey data show that the main reason for vegetation fires is hunting. However, in the total reserve, this activity is formally prohibited, even in the partial reserve, where it is authorized by Order of the Minister in charge of the environment, which specifies the opening and closing dates for hunting in Burkina Faso. As a result, the practice of prohibited hunting in reserves is recurrent.

As villages and fields are located in the reserves, the existence of brushwood fires concerns so-called late fires, as the field preparations in which the fire takes place take place in the locality between March and early May. As the fire cannot be controlled on a field scale, the border areas of the field are where the fire, which was originally intended to prepare the field, burns part of the reserve. The opening up of new fields using traditional fire techniques is also at the origin of late vegetation fires in reserves. Late vegetation fires are also observed in marshy areas around watercourses. Certainly, the herbaceous biomass present in these areas is free from water stress for much of the dry season. These areas are the main refuges for wildlife, and are therefore burned late in the season for hunting purposes.

7.1.4.4. Socio-economic benefits of forest product harvesting

The forestry sector plays an important role in the household economy in terms of the goods and services it provides. Forest products play an active role in generating income for local communities.

7.1.4.4.1. The contribution of forest resources to improving the living conditions of local populations

To claim that forest resources contribute to improving the living conditions of rural households in Burkina Faso would be an understatement. Nonetheless, it would seem wise to determine the areas in which forest products play a role, in order to analyze their real impact on the living conditions of local households.

7.1.4.4.1.1. Wood uses

Wood is used for many purposes:

- ✓ **The use of wood and wood products in culinary and NTFP processing activities**

The main fuel is wood fuel, as almost all respondents (99.) say they use it to cook meals and process forest products, such as shea butter and soumbala (fermented néré seeds). One of the main uses of wood is in the production of sorghum beer. This beverage is made on a number of occasions, including traditional and customary festivals, mutual cultivation, market days and even ordinary days for marketing. The process takes two days, and the finished product is obtained on the third day. The actresses claim to use wood exclusively; some in improved stoves and others in three-stone stoves. In the case of improved stoves, one bundle of wood (photo 12) is enough for one dolo-making session. For three-stone stoves, however, more wood is needed .

Photo12 : Wooden bundles

Field shot (SOME T. Noël, March 27, 2020)

Improved stoves offer an alternative to the rationalization of wood fuel consumption in flourishing commercial activities in the study sites, notably sorghum beer production. Extending the use of improved stoves to all households would considerably reduce the consumption of wood fuel per household, and would ipso facto contribute to the reduction of related deforestation. The work of Kamdem et *al* (2022) in western Cameroon illustrates this point. The study showed that a household's hourly wood consumption is four times higher on a three-stone fireplace than on an improved wood-burning fireplace. Photo 13 shows two improved stoves used to make dolo in the village of Zambo.

Photo13 : Improved fireplaces

Field shot (SOME T. Noël, March 27, 2020)

The fuel used comes from reserves (56.9%), fields (62.) and fallow land (56.), and is either dead wood or green wood cut illegally by the women or purchased. When asked "Are you allowed to take wood from the reserves?", the answer from one of the women was unequivocal: "No, no! We steal to cut wood in the reserves (...)". They also claim to be well-informed about forestry legislation, especially as regards timber harvesting. The origin of fuelwood is not specific to a given area. Wood is harvested wherever it is available, giving rise to illegal practices.

Illegal exploitation of fuelwood causes enough damage to plant formations because no rules in the matter are respected. The negative impact of fuelwood production on vegetation has been reported in Togo's Tchaoudjo Prefecture by Kaina et *al.* (2018). In view of the economic and social importance of this activity, it would be wise for state actors and decentralized communities in these localities to popularize the use of improved stoves in all riparian households in order to minimize wood consumption and by ricochet reduce its impact on the plant formations of these reserves.

Charcoal is used by 35.of respondents, in conjunction with wood. Only 3.and 0.use butane gas and crop residues respectively for cooking. Charcoal comes from the incomplete combustion of firewood during sorghum beer-making activities. Charcoal-making is also carried out clandestinely on reserves by charcoal-makers. It consumes a lot of wood energy, as the production of one kilogram of charcoal requires five kilograms of wood. Hence the harmful impact of this practice on the wood resources of the reserves' plant formations.

The main source of fuel used for cooking and other NTFP processing activities is wood and, to a lesser extent, charcoal. The daily quantities of wood used therefore depend on the number of people per household, their dietary and culinary habits (type of food regularly cooked, storage of food over several days, etc.) and the time of year. The time of peak wood consumption coincides with the rainy season. Food rations, made essentially from leaves, require large quantities of wood. Security stocks of wood built up by women at the start of the dry season are used during this period, when dry wood is scarce and the main players are preoccupied with field work.

- ✓ **The use of wood in housing construction and the manufacture of various materials**

Traditional housing still dominates the survey sites. Traditional dwellings are built using wood as forks, beams and other roofing materials. This observation is attested to by almost all the respondents, for whom the wood of certain plant species is particularly used in the construction of dwellings and sheds because of its quality. *These include Khaya senegalensis, Afzelia africana, Pteleopsis suberosa, Anogeissus leiocarpus, Pterocarpus erinaceus, Azadirachta indica,*

Tectona grandis and Vitellaria paradoxa. Wood is cut from reserves, fields and fallow land. The choice of specific plant species whose wood is deemed to be strong enough to withstand the weight of the terrace and durable (resistant to termite attack) has a definite disadvantage for these species, namely the gradual reduction in their numbers. Some of these species, such as *Khaya senegalensis and Afzelia africana*, are in danger of extinction in the vicinity of the villages, according to the perception of the people interviewed, who regularly mention them.

Green wood is used to make chairs (*Mitragyna inermis*), kitchen utensils (mortars, pestles, spatulas) (*Vitellaria paradoxa* and *Parkia biglobosa*), farming equipment (hoe and machete handles) and other utility tools (ladders, poles, etc.). *Andropogon gayanus* straw is used for roofing certain Mossi and Peulh dwellings, and for sheds in seccos. It is also used in handicrafts to make mats and baskets. Timber harvesting contributes to the reduction of certain plant species insofar as it is selective according to the quality of the wood. Photo 14 shows a ladder made from *vitellaria paradoxa.*

Photo14 : *Vitellaria paradoxa* ladder

Field shot (SOME T. Noël, December 27, 2020)

In customary terms, trees are essential to communities. They are involved in all phases of human life, from birth to death. This is why funeral rites require the use of wood from specific species.

7.1.4.4.1.2. Forest products in household diets

Many NTFPs are fully integrated into the dietary habits of local households. They help manage food shortages and crises, generally during the lean season between June and September, when vulnerable households consume large quantities of these forest products. This is the period when granaries are almost empty and the first harvests are slow in coming (Ouédraogo, 2006). These products, made up of leaves, flowers, fruit, bark and roots, are consumed by 98.5% of those questioned. The use of forest products is likely to increase when crop yields are poor and other sources of income are unavailable (FAO, 2007).

Unprocessed NTFPs are particularly popular during fieldwork. At this time, the unavailability of cereals and the time allotted to women do not allow them to serve several meals (three) daily (Somé, 2012). Hama & *al.* (2019), make a similar observation in the commune of Tamou in south-west Niger in relation to the reasons for NTFP consumption by the populations, which boil down to the fight against hunger in the lean season, the poor agricultural campaign, poverty, the improvement of household living conditions and cultural reasons.

The various plant species and their parts that are consumed are listed in Table 33.

Table33 : Main NTFPs consumed by local households

Botanical name of plant species	Plant parts consumed			
	Leaves	Flowers	Fruits	Nuts/seeds
Vitellaria paradoxa			×	×
Parkia biglobosa			×	×
Adansonia digitata	×	×	×	
Lannea microcarpa			×	
Detarium microcarpum			×	
Saba senegalensis			×	
Balanites aegyptiaca	×	×		
Celtis integrifolia	×			
Tamarindus indica	×	×	×	
Acacia macrostachya	×			
Bombax costatum		×		
Diospyros mespiliformis			×	
Ziziphus mucronata			×	
Gardenia erubescens			×	
Ziziphus mauritiana			×	
Ximenia americana L.			×	
Ficus sycomorus			×	
Annona senegalensis			×	
Vitex madiensis			×	
Crataeva Adansonii D.C.	×			

Source: Field data, March 2020

Fruit is consumed after harvest, without any processing or disinfection, as a snack in stores or at home. This entails the risk of gastrointestinal infections among consumers, especially children, who are most at risk. Leaves, flowers, seeds and nuts, on the other hand, undergo a variety of transformations, from simple drying to cooking followed by drying and preservation. In addition, the preservation of processed NTFPs constitutes a bulwark of food security for households, insofar as it makes condiments permanently available to vary the food ration. As Sanou et *al.* (2022) point out, plants make an important contribution to the diet of riverside communities, as they are sources of various nutrients such as vitamins, calcium, phosphorus and iron.

In short, NTFPs are an integral part of the dietary habits of households in the study area. They therefore contribute to their food security in the sense that they are available in all seasons, and thus help to diversify the food rations of these households. They are part of the diet and nutritional balance of over 43.4% of Burkinabe households (MECV, 2010). Moreover, food security is not exclusively a question of agricultural production. Savadogo (2018) reports regular consumption of NTFPs in other parts of the country, namely Kokologho and Ténado.

7.1.4.4.1.3. Forest products in household health

The health of rural communities living in villages bordering reserves is not just medical; it is largely dependent on forest resources, particularly medicinal plant and herbaceous species. According to WHO (2002), in Burkina Faso, of the population rely on plants for their health care.

Traditional pharmacopoeia still plays an important role in the health of people in this area, as it is used to cure diseases such as malaria, stomach aches, headaches, snake bites and to stabilize chronic illnesses (hypertension and diabetes). Leaves, roots, bark, flowers and fruit are used to treat the various ailments. They are used in decoctions or ground into powder. Application depends on the pathology being treated. A newborn's bath is generally a decoction of leaves, barks and roots of various plant species, to ensure good bodily immunity.

Medicinal plants fall into two broad categories. Plants with a good reputation that are used just about everywhere, a kind of panacea that may be reserved for a particular category of the population; some have very specific properties (Alexandre & Kaire, 2001). The population's recourse to traditional pharmacopoeia lies in its purchasing power, as many households are unable to pay for medical prescriptions and are therefore obliged to turn to nature for the care of their health. However, a lack of confidence in health structures and certain cultural considerations explain why certain categories of the population (the elderly) turn to traditional medicine.

Several studies attest to the health-related uses of medicinal plants by rural populations. This is the case for the uses of *Sclerocarya birrea, Securidaca longepedunculata, Lannea microcarpum and Detarium* reported by Savadogo & *al.* (2019) in the localities of Kokologho and Ténado in the Centre-West region of Burkina Faso. They nevertheless mention that the greater the dependence on and demand for medicinal NTFPs, the greater the possibility of a large impact on the environment. A similar study by Ouattara & *al.* (2016) on floristic diversity and plant use in the Sudanian region of north-west Côte d'Ivoire, characteristic of savannah and open forest formations, argues that the vast majority of the study site's population is dependent on local traditional medicine. They note, however, that the plant parts used are not always communicated. Responding to one of our questions about medicinal plant species, the different parts used in the traditional pharmacopoeia and the pathologies treated, a tradithérapeute from the village of Zoolo asserts "(...) I can't explain that to you! it's a question of genies; it's they who show the use of plants (...)". The therapeutic uses of *Faidherbia albida* and *Khaya senegalensis* by local populations in the Pama partial reserve were also mentioned by Mbayngone & Thiombiano (2011).

Table 34 summarizes some of the medicinal plant species used by people in the study area.

Table34 : Plant species used to treat certain pathologies

Botanical name of plant species	Pathologies treated	Plant parts used
Faidherbia albida	Cough	Barks
Acacia nilotica	Stomach aches	Fruits
Ozoroa insignis	Fever	Leaves
Calotropis procera	Cough, asthma, syphilis, tuberculosis	Leaves, fruit, bark and roots
Piliostigma thonningii	Dermatosis, chickenpox, pimples and skin sores	Leaves
Parkia biglobosa	Treatment of digestive tract pathologies, high blood pressure, parasites	Roots, bark and leaves
Diospyros mespiliformis	Stomach ache, diarrhea and dysentery	Green leaves and fruit
Balanites aegyptiaca	Snake bite, high blood pressure	Leaves and fruit
Vitex madiensis	Fever, malaria and diabetes	Leaves
Ximenia americana	Stomach aches	Roots

Source: Field data, March 2020

These plants are prepared in the form of decoction, juice extraction, maceration, powdering, infusion, calcination (Ky, 2010) prior to administration, which takes place orally (drink or consumption of the product), bath, inhalation, massage, etc.

7.1.4.4.1.4. Forest products as a source of household income

In Burkina Faso, non-timber forest products are the third-largest source of income for rural households at , after agriculture (37%) and livestock products () (OSS, 2015).

It is virtually impossible to quantify the share of forestry income in the total income of households living near reserves, for two main reasons:

- the inorganization of forest management players due to the lack of reserve management with a clear management plan and specifications,

- illiteracy, which makes it difficult to clearly account for financial resources linked to household economic activities.

Moreover, forestry income, while not easily measurable, undoubtedly contributes to overall household income. At least 200 species of trees and shrubs contribute to the diet and treatment of Burkinabè (MECV, 2011). Wood is sold in bundles of 300 F CFA, 500 F CFA, 1,000 F CFA, 1,500 F CFA to households in semi-urban centers; in carts and tricycles to local beer (dolo) processors at prices of up to 10000 F CFA per cartload.

Apart from timber, which is not widely marketed according to the respondents, almost all of whom claim not to be involved in this activity, the bulk of forest income comes from NTFPs, in this case shea kernels and néré seeds. These products are sold either raw or processed.

7.1.4.4.1.5. Income from forest products in other family expenses

The forestry income earned by women, who are the main players in the NTFP and timber sectors, is injected into family expenditure. According to 52.of those surveyed, forestry income is invested in household food (purchase of condiments, foodstuffs), while 21.40% of respondents invest it in the medical care of family members, compared with and 8.respectively, who inject it into the purchase of clothing and kitchen utensils (cooking pots, dishes, etc.), and into children's schooling (purchase of school supplies and payment of school fees). Figure 36 shows the proportions of the different uses of forest income.

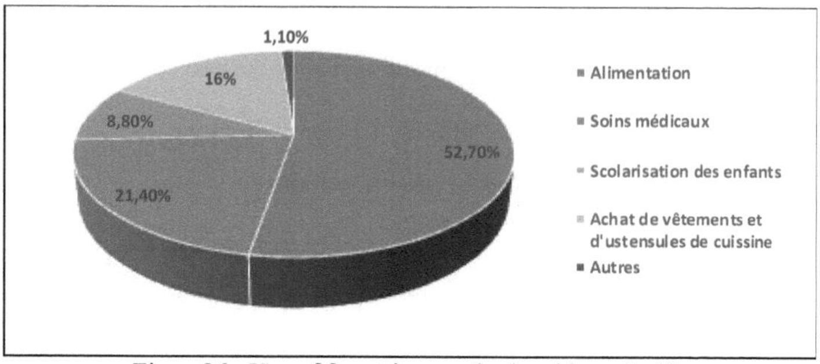

Figure36 : Use of forest income by local households

Source: Field data, March 2020

Forestry income contributes to the overall satisfaction of the vital needs (food and health) of local populations involved in forestry operations. It also contributes to food security, as the main stakeholders have the financial means to purchase foodstuffs, enabling them to stem the food shortage. Food rations could also be varied through condiments to season the sauce and the purchase of foodstuffs that households do not produce.

All in all, the population's needs with regard to the use of forest products are relatively satisfied. Of those surveyed, 52.are very satisfied, while 17.and 8.have a good and average level of satisfaction respectively. The very low and low levels of satisfaction concern 14.15% and 7.of respondents (see table 35).

Table35 : Level of satisfaction of people's needs through the use of forest products

Level of need satisfaction	Workforce	Frequency (%)

Very low	46	14,15
Low	23	7,08
Medium	29	8,92
Good	57	17,54
Very good	170	52,31
Total	325	100

Source: Field data, March 2020

On the whole, people are quite satisfied with the use of forest products to meet their various needs. In fact, these products are used in all spheres of life in these communities. As the exploitation of these products is governed by legislative texts, the populations do not have free access to the products they need in the total reserve. Compensating for the population's needs in terms of forest products should make it necessary and even compulsory to protect the resource.

7.1.5. Local residents' perception of the value of the reserves

Reserves are perceived by local people as a place of worship, a source of food, income and health. The lack of development and structured management of the reserves is not conducive to making the most of this resource. What's more, the main income-generating sectors for households, such as shea, néré, honey and wood, remain informal, despite the existence of a number of women's groups and cooperatives, namely the Groupements Niblossi de Tiankoura, Yembs Taaba Relwendé de Nako and the Union des Coopératives (a group of 14 cooperatives) based in Djikologo, which specialize in the exploitation and processing of NTFPs.

The Union des coopératives, which was created in 2011, has 215 women members. Membership of this structure requires a membership application addressed to the manager, plus the sum of 10,000 F CFA. Membership fees are a source of exclusion for a significant number of women, who are unable to mobilize the funds. The main activities of members are the harvesting and processing of shea nuts and néré sheaths, which adds value to their income. Profits after sales are shared between members of the union. Awareness-raising activities on tree protection and reforestation by members of the cooperative union are organized on a regular basis, according to the person in charge.

With the exception of these few examples of NTFP harvesting and processing organizations, shea kernels and néré seeds are harvested and sold at low prices in their raw state by many players in these sectors. Thc ovcrall low cost of forest products is an obstacle to the emergence of these sectors and, by extension, to the protection of the resource. Surveys and focus groups with women reveal a crucial need to organize the above-mentioned value chains and train the main players, equipping them with appropriate equipment for product processing. Enhancement should be part of a holistic and integrated approach.

7.1.6. Reserves: a place for worship or tourism?

The reserves still abound in sacred sites, potential tourist resources. They are seen as sanctuaries for the worship of divinities by followers of traditional African religion.

7.1.6.1. Traditional cultural practices and the protection of plant and wildlife resources in reserves

The existence of religious practices on reserves remains a reality. It is acknowledged by 63.1% of respondents, while 19.deny it and 17.abstain.

Social and religious organizations emphasize ancestor worship and communion with nature. Sacred woods, hills, trees, rivers, etc. constitute the main altars of worship on which live animals (chickens, sheep, goats and oxen) are sacrificed as offerings to the spirits of ancestors or genies, either for a particular request or to repay a debt. The question of traditional religious rites is becoming increasingly sensitive within these communities, some of whose members who have converted to revealed religions (Christianity and Islam) are reluctant to talk about them for fear of being stigmatized by their new religious communities. All social strata are represented, according to those interviewed. However, to perform a sacrificial rite, the woman is obliged to be accompanied by a man, who may be her brother or her husband. The reason for this, according to the respondents, is that the power of the sacrificer is intended for the man.

The practice of traditional religious rites by local populations represents a bulwark in the protection and management of forest and wildlife resources. All areas dedicated to the worship of supernatural powers are very often protected from human aggression. The cutting of green wood, the gathering of dead wood, pruning, the debarking of certain trees and bushfires are de facto forbidden in sacred places. Their management is therefore based on an approach of conservation and protection through the sacred (Savadogo & *al.,* 2020). These sacred places play a non-negligible ecological role in reserve ecosystems. The positive role in biodiversity preservation played by these sacred natural sites has been mentioned in the Kou classified forest in southwest Burkina Faso by Yaovi (2017). Elsewhere in the country, Sawadogo & *al.* (2017) argue that sacred groves protect several plant species used for food, pharmaceutical purposes and other uses. Above all, they protect species that are becoming rarer (Poda, 1998).

For traditional leaders, the worship of certain age-old trees (*Ceiba pentandra* or the cheese tree and *Adansonia digitata*) is imperative for good rainfall. This story illustrates this perception well: "For three days, we celebrated the funeral of a sacred cheese tree here in the village. Before the third day of the funeral, it rained heavily", revealed a traditional leader from the village of Bontioli. Because of their social organization based on lineage and clan, the ethnic groups settled in the area have totemic prohibitions concerning wild animals, depending on the clan or extended family. Depending on the representativeness of a family or clan in a given environment, one species of wildlife may be conserved to the detriment of others. Nevertheless, it's an organization that helps reduce the depredation of all wildlife. In this area, a fairly representative family (Gbanè) whose totem is the hippotrague, has made it possible to relatively protect the said species in the locality (MEEVCC, 2017).

7.1.6.2. Ecotourism in the reserves

Ecotourism is an almost unknown sector among the populations of the study sites. In response to the question "Do you welcome tourists?", 47.1% of respondents were completely unaware of this activity, compared with 41.who said they did not welcome tourists to their locality. Only 11.1% were in favor of tourists visiting their locality at any given time. Moreover, people carrying out forest research activities are very often likened to tourists by the local population.

Ecotourism is not a preoccupation of the state actors in charge of managing the reserves' forest resources. The current situation in these reserves, characterized by the quasi-permanent presence of man and his various activities, does not lend itself to this type of activity. According to 36.of respondents, the reasons for the very low level of ecotourism in the area are linked to the lack of development of the reserves, and for 32. to the disappearance of wildlife as a result of human pressures. The absence of reception infrastructures and the lack of visibility of the reserves at national and international level in terms of their cultural potential were attested to by 7.and 7.5% of respondents respectively. Unknown reasons were attributed by 19.of respondents.

In any case, the Bontioli wildlife reserves suffer from a lack of vision. Notwithstanding the new challenges posed by gold panning and, above all, insecurity, which mar the State's actions in the overall management of Burkina Faso's forests, developing these reserves and endowing them with a clear management plan, while diversifying activities for the benefit of the communities, could reduce their colonization and attract tourists to the area. All of which could help to create and maintain a well-structured local economy around this resource.

7.2. Forest law enforcement in reserves

Adopted on April 11, 2011, Law N°003-2011/AN on Burkina Faso's Forestry Code replaces the first Forestry Code dated January 31 1997, which lasted just 14 years. The current Code is inspired by the AOF forest regime law. Its objective is clearly defined in Article 1, which states that "the purpose of this Code is to set out the fundamental principles for the sustainable management and development of forest, wildlife and fishery resources". This provision makes special mention of two key principles involved in plant dynamics, namely "sustainability" and "development". These concepts imply individual and collective participation, and substantial investment in forest management by the State for the benefit of the local population. Article 4 paragraph 2 states that the sustainable management of these resources is a duty for all. Without the real participation of the local population, the notion of sustainable forestry will remain an empty word in a context of saturation of rural areas coupled with climatic vicissitudes. This code therefore constitutes a frame of reference through which all development and management actions must be carried out.

7.2.1. Assessing people's level of knowledge of forestry legislation

If forestry legislation is to be applied efficiently, people must first have a very good knowledge of its provisions. From this point of view, an analysis of the mechanisms of its dissemination is necessary.

7.2.1.1. Illiteracy: an obstacle to knowledge of forest legislation?

Formal or informal education contributes to the edification of any society. It is in this context that it is perceived as an important indicator to be taken into account in the process of reserve management and protection by local populations. In this locality, 77.2% of those surveyed claimed to have received no education. Only 4.3% are literate, have primary education, 4.secondary education and 1.higher education.

Statistics from the field survey indicate that most of the rural population in the study area is illiterate. This raises the issue of popularizing innovative techniques for agro-sylvo-pastoral production systems. Ignorance of legislative texts by users of forest resources is an obstacle to good resource management, insofar as they are not translated and explained in the main local languages (Dagara, Lobi and Birifor).

7.2.1.2. Mechanisms for disseminating forest legislation to local communities

The legal principle that "no one is supposed to be ignorant of the law" is applied to the mostly illiterate players involved in forest resource management. There is no formal system for disseminating legislation relating to the environment in general and forestry in particular.

In the field, Eaux et Forêts agents organize a number of awareness-raising sessions on reserve protection. The themes generally addressed revolve around the harmful effects of vegetation fires and the cutting of green wood. The results of the field survey corroborate this. In addition, 67.of respondents maintain that awareness-raising activities are organized by "Forest Watchers" in conjunction with foresters. To ensure that forestry legislation is properly disseminated among the population, one Eaux et Forêts officer asserts: "After an awareness-raising session on a given aspect of forestry management, we assess the level of dissemination of the message by whether the practice continues, decreases or disappears after the awareness-raising session".

However, the difficulty with this technique of disseminating forest legislation based on awareness-raising lies in the credibility enjoyed in their social milieu by the agents charged with carrying out this task and village leaders, and in the sincerity of all the players, from the small number of foresters and "forest supervisors" to carry out the activity regularly. Ouédraogo (2011), attests that one of the major constraints to the appropriation of texts disseminated through the informal circuit, boils down to socio-cultural constraints.

The Forest Code's enforcement mechanism is not very effective either, as prohibited activities are carried out clandestinely in these reserves.

7.2.1.3. Knowledge of plant and animal species protected by forestry legislation

Because of their economic and hunting value, and their importance for biodiversity, certain plant and animal species are protected by current legislation in Burkina Faso. In the villages bordering the Bontioli wildlife reserves, 70.2% of respondents said they were aware of the existence of protected species, compared with 15.who were unaware. On the other hand, 7.4% of respondents did not recognize the existence of such a regulatory provision and 5.abstained.

Farmers seem to have a good knowledge of the regulations concerning the protection of certain species by forestry legislation. In relation to these species, they regularly cite all trees whose fruit is edible and consumable, and in particular *Vitellaria paradoxa, Parkia biglobosa, Adansonia digitata, Lannea microcarpa, Diospyros mespiliformis* and *Saba senegalensis*. With regard to wildlife, a few believe that all wild animals are protected, whereas the vast majority of interviewees regularly mention elephants, lions, panthers and buffalo. As the hunting season is regulated (opening and closing) by the Ministry in charge of wildlife, some farmers maintain that all wild animals are systematically protected by law, which is not the case.

7.2.1.4. Knowledge of prohibited activities in reserves

The Forestry Code of April 4, 2011 prohibits all activities in the total reserve and a certain number of activities in the partial reserve. On the study sites, activities prohibited by forestry legislation are being carried out fraudulently. The local population is well aware of these activities. 54.8% of respondents cited gold panning, 30.late fire, 23.agriculture, 20.other activities including the gathering of immature fruit and the felling of protected trees. Finally, 10.5% of respondents believe that livestock farming is prohibited in the reserves.

Statistics show that local populations, through the awareness-raising sessions organized by Eaux et Forêts agents and "Forest Supervisors", are well aware of the activities they cannot carry out in the partial reserve.

7.2.2. Statutory offences and penalties

Despite the existence of coercive provisions, local populations continue to exploit the reserves, flouting all management rules. According to article 265, the following are punishable by six months' to three years' imprisonment and a fine of ten thousand (10,000) CFA francs to two million (2,000,000) CFA francs, or one of these penalties only:

- those who engage in the commercial exploitation of forest products without prior authorization;
- those who allow animals to roam in forests not open to grazing;
- those who store or transport forest products without authorization;
- those who cut wood without authorization;

- those who introduce exotic plant species into the national territory without authorization;

- those who carbonize forest species in violation of current legislation.

Article 266, paragraph 1, stipulates that anyone carrying out tree lopping or mutilation is liable to a fine of between ten thousand (10,000) and three hundred thousand (300,000) CFA francs.

Article 96 stipulates that: within the boundaries of wildlife reserves and ranches, grazing, clearing, agricultural, forestry or mining activities, waste dumping, polluting activities, uncontrolled fires and, in general, any act incompatible with the conservation and protection of the environment in question, are prohibited. But what are the reasons for this lack of respect for the law?

Reserves such as Bontioli have maintained their status since their creation, without much state intervention to enhance their value, while the populations benefiting from these resources have increased to such an extent that they are succeeding in nibbling away at forest areas. The thorny equation of forest conservation and sustainable management calls for a restructuring of the rural world as a whole, through coherent public policies. The intensification of agro-sylvo-pastoral production, in which the farmer has secure control over his farm, can help local populations to develop a new relationship with their land.

The current Forestry Code has tangible limitations. It does not give enough responsibility to the local population, even if only for the trees grown on their farms. Indeed, producers are inclined to keep the trees on their farms. If they owned these trees, several hundred hectares of land could be regenerated.

Among those responsible for illegal acts in the reserves, 53.8% were responsible for cutting green wood, 15.for gold panning, and 23.for other acts (pruning trees for livestock, debarking trees). Vegetation fires and poaching were cited by only 3.and 3.of respondents respectively (see figure 37).

Figure37 : Prohibited acts committed by the population on reserves

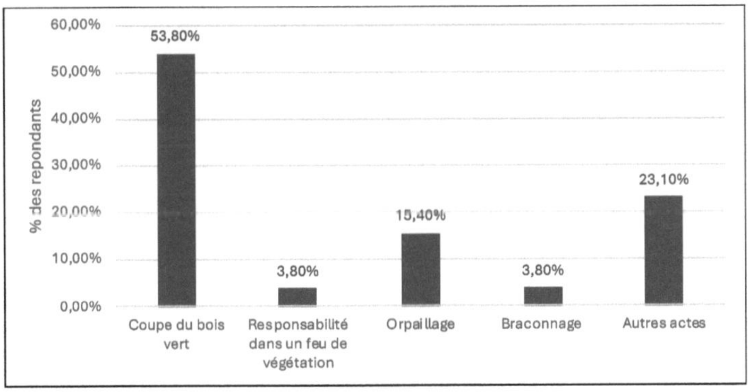

Source: Field data, March 2020

The graph shows that cutting green wood is the main prohibited activity in the reserves. However, this is not an activity that can be carried out in total clandestinity, since when foresters are on patrol, the sound of machete or axe blows guides them. This is why those involved in this practice are easily apprehended by Eaux et Forêts agents. "(...) when they hear the sound of cutting, they spot us, and if we see them from a distance, we throw away our work materials (machetes, cutters) and run away", said a group of women during a focus group. In practice, it is very difficult for foresters to apprehend poachers and those who intentionally start fires. These practices generally take place at night.

In terms of penalties, regulations are not rigorously applied. Those who cut green wood have their work materials (machetes, cutters) confiscated. Those surveyed who have been the victims of financial penalties for carrying out prohibited activities in the reserves estimate the fines paid to the Water and Forestry Department at between 5,000 and 25,000 francs. A number of factors contribute to this state of affairs. The main perpetrators of these reprehensible acts, in this case the women who cut green wood, are unable to face up to the pecuniary penalties provided for under forestry legislation.

Acts such as gold panning and poaching are tainted by corruption at all levels: local, administrative and political. Some orpailleurs use strategies based on corruption to bypass regulations "(...) we often contribute money to give to a forester so that he will let us work", said an orpailleur during an interview. Politicians negotiate the release of local nationals responsible for prohibited acts in the reserves.

However, it must be emphasized that their practices are the antithesis of responsible exploitation of natural resources. Sawadogo (2006) argues that the penalties set out in the Forestry Code are unrealistic given the socio-economic context in which rural populations find themselves, and that the impossibility of paying such penalties provides a breeding ground for corruption.

7.2.4. Difficulties encountered in protecting reserves

The forestry administration, which is responsible for protecting the reserves, encounters various difficulties in carrying out its missions. It lacks the human resources to monitor these protected areas effectively and regularly. In the two main forest posts of Zambo and Djikologo, thirteen (13) Eaux et Forêts agents were counted at the time of the field survey, distributed as follows: four (4) in Zambo and nine (9) in Djikologo, including three (3) trainees. This number is far from sufficient, given the immensity of the tasks assigned to them. These officers do not have enough vehicles (motorcycles) to carry out their patrols. Only two motorcycles were operational. Nor are they adequately equipped to deal not only with poachers, but also with the threat of terrorism, as protected areas are increasingly becoming bases and fallback zones for terrorist groups involved in poaching, mining and other natural resource exploitation. Examples are legion in Africa. Tubiana (2019) attests: "that Boko Haram was able to find refuge in the Sambisa Forest hunting reserve in northeastern Nigeria as well as in the Lake Chad swamps; that W Park is said to have served as a rear base for jihadist attacks in Burkina Faso, that the LRA found refuge in Garamba Park in the DRC and other protected areas in CAR, Sudan and South Sudan; that various armed groups were able to set up in Virunga Park and other protected areas in the DRC, etc. ".

The difficulties associated with protected areas are found throughout sub-Saharan Africa. The major problems of the protected areas system in West Africa are: the weakness of the institutions concerned, the lack of qualified staff and financial resources, obsolete laws and the absence of effective enforcement mechanisms (IUCN/PACO, 2010).

In terms of motivation, there is no longer a Bontioli wildlife reserve surveillance bonus for agents. It has been abolished, and no other form of incentive has been introduced to encourage foresters to protect the reserves.

In addition, the paramilitary corps of Eaux et Forêts is governed by an autonomous status enshrined in LOI N° 063-2015/CNT portant satut du personnel du cadre paramilitaire des Eaux et Forêts in force in Burkina Faso. This law grants specific rights and obligations to these personnel. Article 71 provides that a risk allowance and other allowances to be specified, are granted to paramilitary water and forestry personnel because of the specific and dangerous nature of their functions arising from the duties, missions, attributions, obligations and restrictions of rights it entails.

7.3. Conditions for effective reserve management and recommendations

Biodiversity conservation in the Bontioli wildlife reserves requires that the interests of local populations be taken into account in a participatory, fully inclusive and integrated management plan. Participatory management appears to be a sine qua non for the development of plant formations in these reserves. Indeed, the survival of almost all local populations is intimately conditioned by the exploitation of natural resources (land, water, timber and non-timber products) from these protected areas. With this in mind, Gauthier & Compaoré (2006) argue that it's a good idea to design developments with several objectives in mind: timber, fodder, harvested products and hunting.

FAO (2015) estimates that around 27,000 animal and plant species disappear worldwide every year, due to human activities and certain species. Forests and woodlands are therefore areas of significant biodiversity. To this end, and with a view to restoring biodiversity in all its components in the Bontioli wildlife reserves, and enabling them to perform ecosystem functions for the benefit of the environment and riparian communities, the following actions are proposed:

- ✓ reinforce the forestry service with staff and logistical resources;
- ✓ introduce a financial incentive for Eaux et Forêts agents responsible for monitoring reserves;
- ✓ organize and promote the shea, néré and honey sectors to increase women's incomes, then popularize improved stoves and promote their use through state incentive policies, as well as the use of biogas in all local households;
- ✓ build players' capacities through training;
- ✓ review the status of the partial reserve with a view to its development for participatory and sustainable exploitation of its resources by local populations. To this end, the status could be reviewed in order to clearly define specific units, namely :

- an agricultural production area including housing (this could be done by village). Consensually delimit the fields in the partial reserve with the populations of the administrative villages located there. In order to halt the advance of fields into the plant formations, the development of lowlands for the benefit of farmers living in the reserves and in neighbouring villages would help to significantly reduce harvesting of all kinds and thus limit the fragmentation of plant formations.
- a pastoral zone (grazing area and cattle tracks). The creation of a pastoral zone and cattle tracks would be beneficial for the plant formations in the partial reserve, as illegal and anarchic exploitation is more damaging to the integrity of these plant

formations. By channelling this activity via transhumance trails, we can not only prevent the intrusion of herds into areas not open to grazing, but also rigorously enforce the law with regard to animal roaming in other units.

- an area reserved for wood-energy harvesting for domestic combustion and marketing. This area can be segmented into several harvesting plots.
- a hunting area. This zone will be open to hunting, but with greater protection for certain animal species, especially those in danger of extinction. A complete list of these species could be drawn up by wildlife specialists and made available to all stakeholders, in particular hunters.
- NTFP harvesting could be carried out across all units in strict compliance with a collection code, which would be a guide to techniques and methods for harvesting ripe fruit, leaves, bark, roots, etc., transcribed and disseminated in the main local languages.

Organize farmers into Forest Management Groups (Groupements de Gestion Forestière - GGF), based on the various specificities mentioned above, with legal personality and specifications to be respected. They could be organized by village or by group of villages with close geographical proximity and particular affinities. The GGFs could be coordinated by an umbrella organization called the Union des Groupements de Gestion Forestière (UGGF), whose mission would be to guide, control and assist the GGFs in their operations. As for the GGFs' missions, they could consist of raising awareness and training their members to better exploit and use timber products, non-timber products, grazing, etc. in collaboration with forestry agents, as well as increased surveillance of the reserves. A participatory management plan could be drawn up. As part of a development project, its aim is to involve the community concerned from upstream to downstream. The process is summarized in Table 36

Table36 : Summary of the participatory management plan process

Term	Objective	Tools
Information	Bringing project data and decisions to everyone's attention	Press articles, website, letters or brochures, public meetings, exhibitions, radio announcements in local languages
Consultation	Gather expectations, opinions and views on proposals	Public meetings, sociological surveys, public utility surveys
Consultation	Process of collective discussion between one or more groups to define the actions to be taken. The administrative authority retains the monopoly on final decisions	
Coproduction	Involve people more directly in defining diagnoses and proposals	
Co-decision	Involving local communities in decision-making	Involvement of local communities in steering bodies

Source: Adapted/www.logement.gouv.fr/le-reseau-national-des-amenageurs consulted on 12/26/2024

The value of this approach lies in its inclusive nature, with a view to preventing future conflicts of interest during the implementation of the plan in the field. In the context of the Bontioli wildlife reserves, this plan could concern the following elements:

1- preliminary and total assessment of the plant and wildlife resources of the two reserves ;

2- identification, delimitation and layout of farm plots ;

2- allocation of plots to the GGF ;
3- training of GGF leaders who will be responsible for training all GGF members;
4- putting the delimited units into operation ;
5- opening of paths giving access to plots ;

7- Annual organization of early burning of various plots by GGF members between the end of October and the beginning of November;

8- Annual reforestation organized by GGF members in July and August;

9- organization of permanent surveillance of all units except agricultural production zones by GGF members in collaboration with forestry agents.

As for the total reserve, we suggest that it be granted to a private concessionaire with a view to developing it for hunting and tourism activities, with a view to creating jobs for the local population and protecting all animal species and their habitats. To achieve this, it would be advisable for the State to first develop lowlands to accommodate gold miners operating in the reserve and wishing to convert to agriculture, as well as farmers who were already carrying out their activities there. At the end of this stage, definitively close the gold sites located there (in both reserves) and prohibit any activity in this reserve other than that of the concessionaire.

The management of the Bontioli wildlife reserves has evolved from the traditional approach, aimed exclusively at conservation, towards the involvement of local populations. The intervention of PROGEREF and du PIF has fostered the beginnings of local organization through the creation of peasant structures of hunters, NTFP processors and Forest Management Committees.

Numerous provisions of the Forestry Code govern the exploitation of forest resources in Burkina Faso. Penalties in the form of fines, imprisonment and community service are provided for those who commit acts prohibited by the regulations in force. However, it has to be said that the populations of the study sites have rightly or wrongly invested these protected areas with various activities, thus threatening their biodiversity and causing the fragmentation of their plant formations. The provisions of the Forestry Code are not sufficiently enforced, due to insufficient human resources and equipment. The penalties applied fall far short of those provided for by the regulations. They are not likely to discourage populations in their unhealthy and prohibited practices on forest resources.

Partial conclusion

In the third part of the thesis, it is shown that the occupation of reserves by human settlements and the implementation of extensive agrarian systems by riparian populations have perverse effects on vegetation formations. The cropping system remains unsustainable in the sense that it is mainly based on traditional slash-and-burn techniques and uncontrolled clearing of reserve land for cotton production. The pastoral system also relies exclusively on wild fodder, with pastoral practices that are often unhealthy for the flora.

State officials are finding it difficult to establish an effective management system for the reserves. Nevertheless, the beginnings of participatory management and the organization of the local population are taking shape. The Forestry Code, the main instrument for forest management, clearly defines how forest resources are to be exploited in Burkina Faso. However, demographic and existential factors, as well as technological and technical weaknesses in the fields of agriculture, animal husbandry and energy, have forced farmers to occupy the reserves in an anarchic manner. Incoordination between public policies in the locality is also a source of depredation of plant formations.

GENERAL CONCLUSION

The overall aim of the study was to analyze the evolution and management of plant formations in the Bontioli wildlife reserves. Using a combination of systemic and hypothetico-deductive approaches, the study analyzed, on the one hand, the dynamics of land use and biophysical parameters, in particular the variability of climatic elements on plant formations, by characterizing them. It also examined the influence of agrarian systems and management methods on plant formations.

This work has shown a very marked variation in the different plant formations. Shrub-savanna areas increased considerably in both reserves over the period 1986-2018. This plant formation increased by 43.84% in the total reserve and by 23.07% in the partial reserve. The same is true of bare soils, which also increased from traces to 5.28% and from traces to 26.01% respectively. Fields and fallow land, on the other hand, declined significantly; their areas fell from 60.94% to 0.36% and from 63.39% to 10.80% in the total and partial reserves respectively.

Analysis of the floristic inventory data indicates low generic and specific diversity in all plant formations on both reserves. In the total reserve, 49 species were inventoried, including 36 genera and 18 families. In the partial reserve, 59 species were recorded, including 42 genera and 20 families. Shanonn's diversity indices on the Frontier and Pichod-Viale scales in tropical formations are low overall. They range from 1.78 to 2.68 for plant formations in the total reserve, and from 2.40 to 2.99 for those in the partial reserve. Densities are also generally low.

Nevertheless, they are higher in all plant formations in the total reserve than in the partial reserve. The gallery forests have the highest densities (470 stems/ha for the total reserve and 245 stems/ha for the partial reserve) compared with the other plant formations. The lowest densities are found in agroforestry/ fallow parks. Like densities, basal areas are low in all plant formations on both reserves, with the exception of the gallery forest on the total reserve, which has a basal area of 23 m^2/ha. Structural parameters show high concentrations of stems in the small circumference and height classes.

Analysis of trends in the main climatic parameters over the period 1987-2017 points to an overall deterioration in the climate in this area. Climatic variability is reflected in a number of factors, including falling rainfall. The standardized rainfall index shows years of severe drought (19.35%) and moderate drought (29.03%). Increases in temperature and wind speed were observed over the period 1987-2017. High wind speeds lead to the loosening and lopping of trees. The wind accelerates and amplifies the flames that destroy seedlings, diseased plants, flowers and fruit. Notwithstanding the deterioration of all these climate factors, certain plant formations in both reserves have experienced remarkable growth, notably the shrub savannah, which increased by 43.84% in the total reserve and 23.07% in the partial reserve between 1986 and 2018. Climatic factors do not therefore have a significant influence on the evolution of plant formations. These Sudanian-type plant formations are adapted to this type of climate.

Although field areas have shrunk considerably in the reserves, farmers still practice slash-and-burn cultivation for subsistence and cotton production. The slash-and-burn system, using traditional techniques for opening up new fields and rehabilitating old ones, based on the use of fire and tree-cutting, inevitably leads to fragmentation of plant formations and a reduction in species populations in general, and specifically those of the shrub type. This system of cultivation blocks the evolution of plant formations towards more wooded or even dense stands. Cotton cultivation consumes a great deal of reserve space, and very few plant species are preserved. Selective felling of trees and shrubs followed by ploughing is practiced by most farmers to open up new fields. The pastoral system is entirely dependent on plant resources. Herds stay all year round, and farmers regularly prune plant species such as *Pterocarpus erinaceus, Khaya senegalensis* and *Afzelia africana* to boost livestock feed. The impact of livestock farming on plant formations is marked by heavy use of plant resources by domestic animals and herds of semi-nomadic herders, especially in the dry season.

Management of the Bontioli wildlife reserves has evolved in fits and starts since their creation. At first, it was classic, i.e. oriented exclusively towards conservation. To this end, protection was provided by the State through the Water and Forestry Service. The main activities were surveillance and the repression of prohibited acts. The interventions of PROGEREF and PIF have helped to refocus management through the gradual involvement of the local population, which for the moment concerns only a tiny proportion of the population. This explains why 93% of those surveyed said they were not involved in forest management. The main activities of local communities are not taken into account in the process of managing and conserving these protected areas. The majority of local communities are therefore resistant to conservation. As a result, they continue to practice prohibited activities to a significant extent, such as farming, livestock rearing, gold panning, poaching, vegetation burning and green wood cutting.

In view of these results, the first hypothesis, which states that the plant formations in the reserves have a low floristic richness and diversity and have undergone significant mutations between 1986 and 2018, is confirmed. Despite the deterioration in all the climate elements analyzed, satellite image interpretation shows that some plant formations have progressed. These elements are therefore not significantly correlated with the evolution of plant formations, given their geographical position. The second hypothesis, which stipulates that the main climate parameters have a negative impact on the evolution of reserve plant formations, is therefore partially confirmed. Extensive agro-pastoral activities are practised in the reserves, resulting in the fragmentation of plant formations. This confirms the hypothesis that the inappropriate practices generated by the agrarian systems have had a negative influence on the evolution of plant formations. Furthermore, the low level of local participation in the management of the resources of these protected areas confirms the final hypothesis that local populations are not sufficiently involved in reserve management.

The results of this study open up a number of other avenues for further investigation. These include :

- the correlation between rainfall, temperature and vegetation dynamics in Bontioli wildlife reserves;

- the environmental impact of gold mining on reserves. Gold mining activities pose a serious threat to these protected areas, given their scale and the mining techniques deployed by those involved in the industry;
- characterization of the main NTFP-producing plant species in the Bontioli wildlife reserves, in particular *Vitellaria paradoxa* and *Parkia biglobosa*.

BIBLIOGRAPHICAL REFERENCES

Abba, A. A., Hofs, J.-L., & Mergeai, G. (2006). Meeting environmental challenges for West and Central African cotton sectors. Gembloux, Belgium, *Biotechnol. Agron. Soc. Environ,* v. 10, no. 4, pp. 351-359.

Achard, F., & Blasco, F. (1990). Seasonal rhythms of West African vegetation
by spatial remote sensing. *Remote Sensing and Drought.* AUPELF-UREF ed. John LibbeyEurotext. Paris, France, pp. 1-14.

Adjonou, K., Bellefontaine, R., & Kokou, K. (2009). Les forêts claires du Parc national Oti-Kéran au Nord-Togo: structure, dynamique et impacts des modifications climatiques récentes. *Sécheresse*, v. 20, n° 1e, pp. 1-10.

Affessi, A. S., Koffi, K. G. J. C., & Sangaré, M. (2016). Social and environmental impacts of gold panning on populations in the Bounkani region (Côte d'Ivoire). *European Scientific Journal*, v. 12, no. 26, pp. 288-306.

Ago, E. E. (2016). *Dynamics of carbon fluxes between the atmosphere and West African ecosystems: the case of forests and savannas under Sudanian climate in Benin.* PhD thesis, Université de Liège, 184 p.

Alexandre, D.-Y., & Kaire, M. (2001). Production from African fallows in a Sudanian climate (Wood and miscellaneous products). *Fallowing in tropical Africa.* Ch. Floret, R. Pontanier. John Libbeyn Eurotext. Paris, France, pp. 169-199.

Allies, A. (2018). *Estimation of evapotranspiration by spatial remote sensing in West Africa: towards a better knowledge of this key variable for the region.* PhD thesis in Geosciences, University of Montpellier, 431 p.

Arbonnier, M. (2009). *Arbres, arbustes et lianes des zones sèches d'Afriques* (3ième édition). CIRAD-MNHN-IUCN, Versailles, France, 573 p.

Ariori, S. P., & Ozer, P. (2005). Trends in Soudano-Sahelian West African extent of forest during the second half of the 20th. *Geo-Eco-Trop*, 29, Liège, Belgium, pp. 61-68.

Aubreville, A. (1953). Prospection en chambre XLIII Forêt et climat. *Bois et Forêts des Tropiques*, n°40, Montpellier, France, pp. 63-66.

Aubreville, A. (1957). Yangambi agreement on the nomenclature of African vegetation types. *Bois et Forêts des Tropiques,* v. 51, pp. 23-27.

Aurouet, A., Devineau, J.-L., & Vidal, M. (2005). Les principaux facteurs de l'évolution des milieux riverains du Mouhoun près de Boromo (Burkina Faso): changement climatique ou dégradation anthropique? *Sécheresse*, v. 16, no. 3, pp. 199-207.

Avana-Tientcheu, M. L. A., Keouna, S., Dongock Nguemo, D., & Mouga Masdewel, B. (2019). Stand structure and domestication potential of *Parkia biglobosa* in the Tandjilé-Ouest region (Chad). *International Journal of Biological and Chemical Sciences,* v. 13, no. 1, University of Dschang, Cameroon, pp. 219-236.

Avakoudjo, J., Mama, A., Toko, I., Kindomihou V., & Sinsin, B. (2014). Land use dynamics in the W National Park and its periphery in northwest Benin. *International Journal of Biological and Chemical Sciences*, v. 8, no. 6, University of Dschang, Cameroon, pp. 2608-2625.

Bakouan, N. J. (1995). *La gestion décentralisée des ressources forestières: cas du projet BKF/89/011,* Ouagadougou, 103 p.

Ballouche, A., & Rasse, M. (2007). L'homme, artisan des paysages de savane. *Pour la Science,* n° 358, pp. 56-61.

Bamba, I. (2010). *Anthropization and spatio-temporal dynamics of forest landscapes in the Democratic Republic of Congo*. PhD thesis, Université Libre de Bruxelles and Université d'Europe, 205 p.

Bandré, E. (1996). Anthropisation du couvert végétal dans la province du Namentenga et ses conséquences socio-économiques: cas de la région de Tougouri en zone sub-sahélienne (centre-nord du Burkina Faso). Proceedings, *Berichte des Sonderforschungsbereichs 268, Band 7, Frankfurt a.M.,* pp. 17-31.

Barbara, L. (2012). *Guidelines for protected area legislation.* Gland, Switzerland: IUCN. xxviii + 406 p.

Beaux, J.-F. (2011). *L'environnement.* 3e edition, Nathan, Paris, 159 p.

Bellefontaine, R., Sabir, M., Kokou, K., Guinko, S., Saadou, M., Ichaou, A., Hatem, C., Bationo, B. A., Karim, S., & Logbo, J. (2003). *Revegetation, a fourth way: natural vegetative propagation*. XII World Forestry Congress (Montreal), 8 p.

Bergaoui, M. & Alouini, A. (2001). Characterization of meteorological and hydrological drought: the case of the Siliana watershed in Tunisia. *Sècheresse*, v. 12, no. 4, pp. 205-213.

Bertrand, A. (1985). Les nouvelles politiques de foresterie en milieu rural au Sahel: Règlementations foncières et forestières et gestion des ressources ligneuses naturelles dans les pays de la zone soudano-sahélienne. *Revue Bois et Forêts des Tropiques*, n° 207, 1er trimestre 1985, Montpellier, France, pp. 23-39.

Bertrand, G. (1968). Landscape and global physical geography. Esquisse méthodologique. *Revue Géographique des Pyrénées et du Sud-Ouest*, fasc. 39, Toulouse, France, pp. 249-272.

Bigot, S. (2004). *Climate variability, interactions and environmental modifications. L'exemple de la Côte d'Ivoire.* Habilitation à Diriger des Recherches, Université des Sciences et Technologies de Lilles and Centre National de Recherche Scientifique, 398 p.

Birouk, A. *La convention sur la diversité biologique et sa mise en œuvre au Maroc.* Institut Agronomique et Vétérinaire Hassan II, Rabat, 31 p. https://www.thegef.org

Bonfils, M. (1987). *Halte à la désertification au Sahel.* Editions KARTHALA, Paris, France, 263 p.

Bognounou, F. (2004). *Characterization and management of forage ligneous plants in agro-pastoral production systems in the Dankana terroir in the South Sudanian zone of Burkina Faso.* Mémoire de D.E.A en Sciences Biologiques, Université de Ouagadougou, 87 p.

Bognounou, F., Thiombiano, A., Savadogo, P., Boussim, J. I., Oden, P., & Guinko, S. (2009). Woody vegetation structure and composition at four sites along latitudinal gradient in Western Burkina Faso. *Bois et Forêts des Tropiques*, v. 300, no. 2, pp. 29-44.

Bonou, A. (2008). *Estimation de la valeur économique des Produits Forestiers Non Ligneux d'origine végétale dans le village de Sampéto (commune de Banikoara).* Dissertation for DEA in Aménagement et Gestion des Ressources Naturelles (AGRN), Université d'Abomey - Calavi (Bénin), 66 p.

Bordes, C. (2010). *La gestion des arbres par les paysans: étude d'une enclave au milieu de réserves forestières au sud-est du Burkina Faso.* Final dissertation, Ecole d'Ingénierie en Agro-Développement International, ISTOM, Cergy-Pontoise Cedex, 110 p.

Boserup, E. (1972). *Conditions of agricultural growth.* Reply to Sheffer, American Antiquity, 37, London, 447 p.

Botoni, E., & Reij, C. (2009). *The silent transformation of the environment and production systems in the Sahel: impacts of public and private investments in natural resource management.* Center for International Cooperation (CIS), Vrije Universiteit amsterdam, Comité Inter-Etat de Lutte contre la Sécheresse dans le Sahel (CILSS), Ouagadougou, 63 p.

Boubacar, H. (2010). *Biophysical characterization of woody resources in degraded and revegetated areas of the Sahel: case of the Mayahi department.* Mémoire de D.E.A en Sciences et Techniques, Université Abdou Moumouni, Niamey, 66 p.

Bouda, Z. H.-N. (2008). Textes et textes de loi sur la gestion des ressources naturelles au Burkina Faso, CIFOR, Ouagadougou, 160 p.

Boudrais J. (2008). Pastoralism and protected areas in West Africa and East Africa. In Aubertin C. & Rodary E. (Eds.). *Aires protégées, espaces durables?* IRD Éditions, Marseilles, pp. 215-246.

Bonko, B. S., Sinsin, B., & Soule, B. G. (2007). Effet de la dynamique d'occupation du sol sur la structure de la diversité floristique des forêts claires et savanes au Bénin, *TROPICULTURA*, v. 25, n° 4, Liège, Belgique, pp. 221-227 .

Brown, L. R. (2007). *Le plan B: pour un pacte écologique mondial,* Calmann-Lévy-Souffle Court, HACHETTE Littérature, Paris, 415 p.

Broussier, J. (1987). System and production system. *Cah. Sci. Hum.* v. 23, no. 3-4, Paris, pp. 377-390.

Buttoud, G., & Samyn, J.-M. (1999). *Politique et planification forestière.* Guide pour la formulation et l'élaboration, Intercoopération, 87 p.

Cabala Kaleba, S., Useni Sikuzani, Y., Sambieni K. R., Bogaert, J., & Munyemba Kankumbi, F. (2017). Dynamics of the forest ecosystems of the Katangese Cuprifere Arc in the Democratic Republic of Congo. Causes, spatial transformations and magnitude. *TROPICULTURA*, v. 35, no. 3, pp. 192-202.

Calvet, R., Burriuso, E., Bedos, C., Benoît, P., Charnay, M.-P., & Coquet, Y. (2005). *Pesticides in the soil: agronomic and environmental consequences*. Editions France Agricole, Paris, 634 p.

Caillaut, S., Ballouche, A., & Delahaye, D. (2009). Spatio-temporal organization of bushfires. A comparative approach in Burkina Faso. Neuvièmes Rencontres de Théo Quant Besançon, March 4-6, 2009, 10 p.

Cambrézy, L., & Sangli, G. (2011). The geographical effects of population growth in rural Africa: the example of Southwest Burkina Faso. *CFC*, n° 207, Brussels, Belgium, pp. 75-93.

Cecca, F., & Bagils, R. Z. (2015). Paleobiogeography. Collection Géosphères, *EDP Sciences*, Les Ulis, France, 22 p.

Chardonnet, B. (2019). *Africa is changing: should its protected areas evolve? Reconfiguring protected areas in Africa*. IUCN/Papaco, 46 p.

Chidumayo, E., Okali, D., Kowero, G., & Larwanoum, (Eds.). (2011). *Forests, wildlife and climate change in Africa*. African Forest Forum, Nairobi, Kenya, 356 p.

CILSS. (2002). *Sahel: natural resources, keys to development*! Ouagadougou, 28 p.

CILSS. (2010). *Le Sahel face aux changements climatiques : Enjeux pour un développement durable*. Center Régional AGRHYMET, Bulletin Mensuel, Numéro spécial, Ouagadougou, 43 p.

Cochet, H. (2007). Innovation capacity of farming systems and natural resource management in Burundi. *AgroParisTech*, 19 p.

Cochet, H. (2011). Origine et actualité du " système agraire " : retour sur un concept. *Revue Tiers Monde,* n° 207, Montreuil, pp. 97-114.

Cohen, M. (2008). La biogéographie à l'épreuve de l'interdisciplinarité : proposition d'une démarche de recherche. French version of article published in the Proceedings of the Séminaire International *Interdisciplinariedade e Universidade noseculo XXI*, UMR Ladyss-Université Paris Diderot, 12 p.

CPCS. (1967). *Soil classification*. Ed. ENSA, Grignon, 100 p.

Dabiré, B. H. (2016). *Migration in Burkina Faso: migration profile 2016*. International Organization for Migration (IOM), Geneva, Switzerland, 154 pp.

Da, D. E. C. (1984). *Recherches géomorphologiques dans le Sud-Ouest de la Haute Volta: la dynamique actuelle en pays Lobi*. Thèse de doctorat de troisième cycle de Géographie, Université Louis Pasteur, Strasbourg, 349 p.

Da, D. E. C. (1996). La nature sera toujours plus forte que l'Homme. *Revue Pyramide*, n° 00, Ouagadougou, pp. 43-46.

Da, D. E. C. (2005). From systems approach to geographic information systems. Proceedings of the 2th international symposium on *"Remote sensing and geographic information systems (GIS). Experience and prospects for development-oriented university research and training"*. Ouagadougou, January 18-19, 2005. Cahier du CERLESH , Special issue, pp. 21-36.

Da, D. E. C., & Yonkeu, S. (2007). *Definitions, concepts et objectifs des études en gestion des ressources et environnement*; 2iE, Université de Ouagadougou, PowerPoint.

Daou, I., Coulibaly, A., Sidibé, A., Sangaré, H., Kéita, I., Bolozogola, Y., Touré, A., & Mariko, A. (2019). Suivi de la dynamique environnementale de 1985 à 2018 en zone soudano-sahélienne par télédétection: cas de la commune rurale de Nyamina. *Rev. Ivoir. Sci. Technol.* n° 34, pp. 13-24.

De Foresta, H. (2008). Forests and forestry in tropical regions. In Hallé, F., & Lieutaghi, P. (Eds.) *Aux origines plantes*, v. 2, Editions Fayard, pp. 138-155.

Defourni, P. (2017). Mapping and monitoring land cover. In Delince (ed.), *Handbook on the application of remote sensing to agricultural statistics (chapter 2).* Handbook of the Global Strategy for Improving Agricultural and Rural Statistics (GSARS): Rome, pp. 21-59.

Deleporte, P. (2002). Introduction: biogeography and systematics, problems and methods. *Biosystema 20, Systématique et biogéographie*, Paris, pp. 3-6.

Demont, M., & Jouve, P. (1999). Evolution of village agro-ecosystems in the Korhogo region (northern Côte d'Ivoire): Boserup versus Malthus, opposition or complementarity. *Agrarian dynamics and the social construction of territory.* Séminaire CNEAR-UTM, 26-28/04/1999, Montpellier, France, pp. 93-108.

Déry, S. (2008). Protected areas, a new tool for the dynamism of Southeast Asian states? *ANNALES DE GÉOGRAPHIE*, v. 1, n° 659, pp. 72-92.

Descoings, B. (1970). Les flores régionales: méthodologie de l'inventaire floristique. *Naturalia monspeliensia*, série Bot. 21, Centre d'Etudes Phytosociologiques et Ecologiques, Montpellier, pp. 51-73.

Descoings, B. (1978). Les formations herbeuses dans la classification phytogéographique de Yangambi. *Adansonia*, ser. 2, v. 18, no. 2, Paris, pp. 243-256.

Devineau, J.-L. (1997). Évolution saisonnière et taux d'accroissement des surfaces terrières des ligneux dans quelques peuplements savanicoles soudaniens de l'ouest burkinabé. *Ecologie*, v. t. 28, n° 3, pp. 217-232.

Devineau, J.-L., & Fournier, A. (1998). *Écologie d'une savane africaine: Synthèse provisoire des résultats acquis.* ORSTOM/ERMES, Orleans, France, 77 p.

Dialla, B. E. (2003). La question foncière sur les périmètres hydro-agricoles du Burkina Faso. Center d'Analyse des Politiques Économiques. *SERIE DOCUMENT DE TRAVAIL DT-CAPES N° 2003-07*, Ouagadougou, 24 p.

Diallo, B. A., Diarra, B., Touré, M., Cissé, D. A., & Doumbia, B. (2020). Urban sprawl in Bamako: explanatory factors and implications. *Afrique SCIENCE,* v. 17, no. 6, pp. 58-75.

Diallo, M. S. (2011). *Evolution of protected area management in Guinea: the difficult cohabitation of public policies and traditional systems: the case of the Haut Niger National Park*. Doctoral thesis in Geography, Université du Maine, 269 p.

Diarra, A. (1999). *Étude d'un modèle d'aménagement des forêts naturelles en zone soudano-sahélienne au Burkina Faso*. Master of Science thesis, Faculty of Forestry and Geomantics, Université Laval, Canada, 101 p.

Diédhiou, M. A. A., Faye, E., Ngom, D., & Fall, S. (2018). Characterization of the flora and woody vegetation of the village terroirs of Keur Birame (Kaffrine) and Saré Yorobana (Kolda) in Senegal. *European Scientific Journal,* July 2018 edition, v. 14, n° 21, Quebec, Canada, pp. 391-409.

Diello, P., Mahe, G., Dezetter, A., Delclaux, F., Servat, E., & Ouattara, F. (2005). Vegetation index-rainfall relationships in Burkina Faso: the case of the Nakambé watershed. *Hydrological Sciences-Journal-des Sciences Hydrologiques*, v. 50, no. 2, Dakar, Senegal, pp. 207-221.

Dimobé, K. (2017). *Dynamics, carbon sequestration and patterns of variation in Sudanian savannas of Burkina Faso and Ghana (West Africa)*. PhD thesis in Applied Biological Sciences, Université de Ouaga I Pr Joseph KI-ZERBO, 207 p.

Diouf, M., Nongierma, A., Amani, A., & Somé, A. R. B. (2000). Drought control in the Sahel: results, achievements and prospects at the AGRHYMET regional center. *Sciences et changements planétaire/Sécheresse*, v. 11, no. 4, Niamey, Niger, pp. 257-266.

Dipama, J.-M. (1997). *Les impacts du Barrage hydroélectrique sur le bassin versant de la Kompienga (Burkina Faso)*. PhD thesis, University of Bordeaux III, 392 p.

Dipama, J.-M. (2009). Ecosystem degradation risks associated with cotton cultivation in Burkina Faso: the case of Pô National Park. *Revue de l'Université de Moncton,* v. 40, no. 2, pp. 29-52.

Djego, J., Gibigaye, M., Tente, B., Sinsin, B. (2012). Ecological and structural analysis of the Kaodji community forest in Benin. *International Journal and Biological and Chemical Sciences*, v. 6, no. 2, University of Dschang, Cameroon, pp. 705-713.

Doamba, P. (2012). *Impact of the use of timber and non-timber forest products on the management of Arly National Park in Burkina Faso*. Master's thesis in environment, Institut International d'Ingénierie de l'Eau et de l'Environnement, Ouagadougou, 49 p.

Dorsouma, A.-H. & Mélanie, R.-D. (2008). Climate variability, desertification and biodiversity in Africa: adapting, an integrated approach. *Vertigo -la revue électronique en science de l'environnement* [online], v. 8, n° 1, URL : http://journals.openedition.org/5356/vertigo

Doumenge, C. (2021). Protecting biodiversity: a look back at the evolution of "protected areas" around the world. *CIRAD*, 6 p.

Drouin, J.-M. (1995). La biogéographie : modèle unique ou histoire commune ? In Chatelin, Y., & Bonneuil, C. (Éds.). *Les sciences hors d'Occident au XXe siècle,* Nature et Environnement, v. 3, Paris, pp. 129-140.

Dudley, N. (Editor) (2008). *Guidelines for applying management categories to protected areas.* Gland, Switzerland: IUCN, 96 p.

Dumont, G.-F. (2014). Population and nature: antagonism or concordance? *Sustainable Human, Sustainable Nature: Our Responsabilit*, Vatican City, 36 p.

Dykstra, D. P., Koxero, G. S., Ofosu-Asiédu, A., & Kio, P. (1996). *Promoting forest management in the tropical rainforest zone of English-speaking West and Central Africa.* United Nations Environment Programme (UNEP) and Center for International Forestry Research (CIFOR), Jakarta, Indonesia, 125 p.

Etzold, S., Wunder, J., Braun, S., Rohner, B., Bigler, C., Abegg, M., & Rigling, A. (2016.) Forest tree mortality: causes and trends. In Pluess, A., Augustin, S., & Bang P. (Eds.), *Forests and climate change. Elements for adaptation strategies.* Federal Office for the Environment, FOEN, Bern, Federal Research Institute WSL, Birmensdorf, Haup; Bern, Stuttgart, Vienna, pp. 183-202.

FAO (2001). Global forest resources assessment 2000. *Main report.FAO Forestry paper,* (140), Rome, pp. 115-120.

FAO (2007). Forests and sustainable livelihoods: what role can forests and forestry play in poverty reduction? *Unasylva* 226/227, v. 58, Rome, pp. 80-87.

FAO (2008). *Overview of the national fisheries sector in Burkina Faso.* Ouagadougou, 19 p.

FAO (2010). *Global Forest Resources Assessment 2010: terms and definitions.* Rome, 30 p.

FAO (2011). *State of the World's Forests.* Rome, 176 p.

FAO (2015). *Global Forest Resources Assessment 2015.* Rome, 253 p.

FAO (2018). *Africa on the move. Dynamics and factors of migration south of the Sahara.* Rome, 58 p.

FAO (2020). *Global Forest Resources Assessment 2020: Main findings.* Rome, 12 p.

Fourgeau, C. (2002). Les petites jachères des femmes bwaba et bobo: perception et définition identitaire. Provinces of Balé and Houet, Burkina Faso. In Lacombe, B., Palé, F. O. K., & Traoré, S. (Eds.). *Études sur la jachère dans le Sud-Ouest du Burkina Faso: contribution des sciences sociales à la définition des relations sociétés rurales avec les jachères,* pp. 136-153.

Fournier, A. (1991). *Plant phenology, growth and production in some West African savannas: variation along a climatic gradient.* Editions de l'ORSTOM, Collection ÉTUDES et THÈSES, Paris, 314 p.

Fournier, A., Floret, C. & Gnahoua, G.-M. (2001). Fallow vegetation and post-cultivation succession in tropical Africa. *John Libbey Eurotext,* Paris, pp. 123-168.

Fournier, A., & Sinsin, B. (2007). Les aires protégées d'Afrique de l'Ouest, une identité en devenir? In Fournier, A., Sinsin, B., Mensah, G. A. (Eds.). *Quelles aires protégées pour l'Afrique de l'Ouest? Biodiversity conservation and development.* Collection Colloques et Séminaires, Paris, IRD, pp. 28-40.

Fournier, A., & Millogo-Rosolodimby, J. (2007). Vegetation threatened or shaped by man? In Fournier, A., Sinsin, B., Mensah, G. A. (Eds.). *Quelles aires protégées pour l'Afrique de l'Ouest? Conservation de la biodiversité et développement.* Collection Colloques et Séminaires, IRD, Paris, pp. 62-76.

Fournier, A., Douanio, M., & Béné, A. (2012). Practice and perception of vegetation fires in an orchard landscape. Le pays sèmè (Kénédougou, Burkina Faso). Colloque de Ouagadougou 8-10 mars 2012, 31 p.

Frontier, S., & Pichod-Viale, D. (1991). *Ecosystèmes: structure, fonctionnement, évolution.* Paris, Masson. Collection d'écologie 21, 392 p.

Gansaonré, R. N., Sodoré, A. A., & Ouédraogo, B. (2020). Jeu des acteurs à la périphérie du Parc W du Burkina Faso: entre conflictualité et incoordination des interventions. *Afrique SCIENCE*, v. 16, no. 5, Abidjan, Côte d'Ivoire, pp. 118-135.

Gautier, D., & Compaoré, A. (2006). *Les populations locales face aux normes d'aménagement forestier en Afrique de l'Ouest. Mise en débat à partir du cas du Burkina Faso et du Mali.* CIFOR, West Africa Regional Office, Ouagadougou, Burkina Faso, 63 p.

Gautier, D., Ouédraogo, G. G., Badini, Z., & Dabat, M.-H. (2009). *Appui à la définition de stratégies de développement des filières agro-sylvo-pastorales et halieutiques sélectionnées dans les régions d'intervention du PADAB II " Goulots d'étranglement et actions pilotes "*, CIRAD, Rapport filière bois, Région du Centre-est, Ouagadougou, 57 p.

George, P., & Verger, F. (2009). Dictionnaire de géographie, 10(ème) édition, QUADRIGE/PUF, Paris, 512 p.

Giazzi, F., & Tchamie, T. T. K. (2007). La participation des populations dans la gestion des ressources naturelles: historique de la création des aires protégées en Afrique de l'Ouest et évolution récente de la conservation. In Fournier, A., Sinsin, B., Mensah, G. A. (Eds.). *Quelles aires protégées pour l'Afrique de l'Ouest? Conservation de la biodiversité et développement.* Collection Colloques et Séminaires, IRD, Paris, pp. 77-86.

IPCC. (2000). *Land use, land-use change and forestry.* Summary for decision-makers. Special Report of the Intergovernmental Panel on Climate Change, Geneva, Switzerland, 30 p.

Giezendanner, F. D. (2012). Random sample size and Margin of Error. CMS-SPIP, Geneva, 22 p. http://icp.ge.ch/sem/cms-spip/spip.php?article1641

Giraut, F., Guyot, S., & Houssay-Holzschuch, M. (2004). Protected areas in African territorial recompositions. *L'Information Géographique*, v. 68, n° 4, Paris, pp. 340-368.

Gnoumou, A., Ouattara, H. A., Sambaré, O., & Ouédraogo, A. (2021). Characterization of the diversity and structure of woody vegetation in riparian formations of the Kari classified forest, Burkina Faso. *Afrique SCIENCE*, v. 18, no. 1, Abidjan, Côte d'Ivoire, pp. 69-89.

Gomgnimbou, M. (1998). Forêts à identité contestée: perception des classements des forêts de Tiogo et de Laba pour les populations locales. *Aménagement intégré des forêts naturelles des zones tropicales sèches de l'Afrique de l'Ouest.* Proceedings of the International Seminar, November 16-20, 1998, Ouagadougou, Burkina Faso, pp. 233-241 .

Gomgnimbou, A. P. K., Savadogo, P. W. W., Nianogo, A. J., & Millogo-RAasolodimby, J. (2010). Farming practices and farmers' perceptions of the environmental impacts of monoculture in Kompienga province (Burkina Faso). *Science-Nature*, v. 7, n° 2, Ouagadougou, pp. 165-175 .

Goudet, J.-P. (1985). Natural forest management and reforestation in tropical dry Africa. *Revue Aménagement et Nature,* n° 81, pp. 33-36 .

Gramond, D. (2010). *Introduction to biogeography.* CNRS, Université Paris-Sorbonne, 28 p.

Guéneau, S. (2011). *Vers une évaluation de dispositif de prise en charge du problème de déclin des forêts tropicales humides.* PhD thesis, Institut des Sciences et de l'Industrie du Vivant et de l'Environnement (AgroParisTech), 581 p.

Guillard, J. (2014). *Au service des forêts tropicales: histoires des services forestiers français d'outre-mer 1896-1960.* Tome 1, AgroParisTech, Centre de Nancy, 641 p.

Guinko, S. (1984). *Végétation de la Haute- Volta.* Thèse de doctorat d'Etat en Sciences Naturelles, Université de Bordeaux III, Tome 1, 318 p.

Guira, M. (1997). *Etude de la phénologie et de la variété de quelques caractères chez le karité Butyrospermum paradoxum sub sp. Parkii (G. Don) Hepper (Sapotaceae) dans les champs et les jeunes jachères dans la moitié ouest du Burkina Faso.* Thèse de doctorat de Troisième Cycle en Biologie et Ecologie Végétale, Université de Ouagadougou, 191 p.

Guira, M. (1997). *Contribution à l'étude de la dégradation du couvert végétal à Karangasso-Vigué.* Master's thesis in Geography, University of Ouagadougou, 113 p.

Hadef, A. (2010). *Vegetation land use mapping from satellite data in the Annaba region (Chétaibi).* Master's thesis in Plant Ecology, Université Badji Mokhtar Annaba, 115 p.

Hama, O., Tinni, I., & Baragé, M. (2019). Diversity and importance of non-timber forest products of plant origin in the rural commune of Tamou, southwest Niger, West Africa. *Rev. Ivoir. Sci. Technol.* n° 34, Abidjan, Côte d'Ivoire, pp. 191-215 .

Hauchart, V. (2005). *Cotton cultivation and soil degradation in Mouhoun (Burkina Faso).* Doctoral thesis in Geography, Université de Reims-Champagne-Ardenne, 467 p.

Hema, K. B., Zerbo, R., & Koffi, G. R. Y. (2018). Agricultural practices of Burkinabè migrants and socio-economic investments related to cocoa cultivation. *TROPICULTURA*, v. 36, no. 2, Liège, Belgium, pp. 299-313 .

Hien, P. C. (2000). Frontiers and conflicts among the Dagara and their neighbors in southwest Burkina Faso (XVIIIème-XIXème siècle). *Berichte des Sonderforschungsbereichs 268, Band 14, Frankfurt a.M.*, pp. 427-440 .

Hountondji, Y. C. H. (2008). *Dynamique environnementale en zone sahélienne et soudanienne d'Afrique de l'Ouest : Analyse des modifications et évaluation de la dégradation du couvert végétal.* PhD thesis in Environmental Sciences and Management, Université de Liège, 153 p.

Hugot, J.-P. (2002). Historical biogeography: the study of biological diversity in space and time. *Biosystema 20, Systématique et biogéographie*, Paris, pp. 27-40.

Ibrah, H. (2005). *Produits forestiers non ligneux et gestion locale des ressources naturelles: cas de département de Filingué au Niger.* DEA in Environmental Management, Senghor University, Alexandria, Egypt, 82 p.

Ibrahim, B. (2012). *Characterization of rainy seasons in Burkina Faso in a context of climate change and assessment of hydrological impacts on the Nakambé basin.* PhD thesis in Earth Sciences, Université Pierre et Marie Curie and Institut International d'Ingénierie de l'Eau et de l'Environnement (2iE) de Ouagadougou, 246 p.

INSD. (2009). *Projection démographique de 2007 à 2020 par région et par province.* Ouagadougou, 69 p.

Isnard, H. (1952). Le mécanisme du climat de l'Ouest-africain. *L'information géographique*, 16, 3, Paris, pp. 98-100.

Issifou Moumouni, Y., Toko Imorou, I., & Moussa, A. D. (2019). Characterization of degradation through floristic diversity and vegetation structure in the middle Sota basin of North Benin. *International Journal of Progressive Sciences and Technologies (IJPSAT),* v. 18, n°1, pp. 53-70.

Jacquemot, P. (2018). What future for Africa's protected areas? *Dossier Willagri*, 23 p.

Jiagho, E. R., Zapfack, L., Banoho, L. P. R. K., Tsayem-Demaze, M., Corbonnois, J., & Tchawa, P. (2016). Diversity of woody flora at the periphery of Waza National Park (Cameroon). *Vertigo. La revue électronique des sciences de l'environnement,* v. 16, n° 1, 56 p. revues.org

Job J.-O. (1992). Les premiers aménagements hydroagricoles en Mésopotamie et les problèmes de sels. *ARCTIQUES/ORSTOM* Editions, Paris, pp. 473-480.

Jouve, P. (2004). Population growth, brake or opportunity for sustainable agricultural intensification in Africa. *Transition agraire et résilience des sociétés 'rural', courrier de l'environnement de l'INRA,* n°52, CNEARC, Montpellier, pp. 101-106.

Jyrki, J., Vanha, I., & Tonteri, T. (1998). Optimal Sample and plot size for inventory of field and ground layer vegetation in a mature Myrtillus-type boreal spruce forest. *Ann. Bot. Fennici*, v. 35, pp. 191-196 .

Kaboré, A. (2010). *Community strategies for adaptation to climate change: the case of sacred woods in the Moaaga socio-cultural area of Burkina Faso*. Doctoral thesis in Geography, Université d'Abomey-Calavi, 225 p.

Kaboré, A. (2010). *One man's bush, another man's protected area. Histoire du peuplement, perception de la nature et politique des aires protégées dans le Gourma au Burkina Faso: l'exemple de la Réserve partielle de faune de Pama.* PhD thesis in development studies, Graduate Institute of International and Development Studies, Geneva, 392 p.

Kaboré, O. (2013). *Land use dynamics in savanna ecosystems and agrarian system of the Kompienga watershed (Burkina Faso)*. PhD thesis in Geography, University of Ouagadougou, 281 p.

Kaboré, O., Ouédraogo, B., & Sawadogo, I. (2019). Fragmentation factors and management strategies of forest massifs in Burkina Faso. *Journal of Applied Biosciences*, v. 133, Nairobi, Kenya, pp. 13516-13531.

Kaboré, S. (2018). *Contribution de l'agroforesterie à Faidherbia albida dans l'adaptation paysanne à la variabilité climatique à Imasgo (dans la province du Boulkiemdé, Burkina Faso)*. Master's research thesis in Geography, Université Ouaga I Pr Joseph KI-ZERBO, 95 p.

Kadéba, A., Kagambèga, F. W., Soulama, S., Schmidt, M., Thiombiano, A., & Boussim, J. I. (2019). Dynamics of vegetation units and response of woody species to land degradation in the sub-Sahelian zone of Burkina Faso. *Rev. Ivoir. Sci. Technol.* n° 34, pp. 144-162.

Kadjo, B. (2010). *Aménagement en techniques de conservation des eaux et des sols du bassin versant du Ioba (Dano) en vue du reboisement et de la production améliorée du Jatropha Curcas: quelles options optimisées?* Master's thesis, Institut International d'Ingénierie de l'Eau et de l'Environnement (2iE), Ouagadougou, 73 p.

Kaina, A., Wala, K., Koumantiga, D., Folega, F., & Akpagana, K. (2018). Impact of wood energy exploitation on vegetation in the Prefecture of Tchaoudjo, Togo. *Revue de Géographie de l'Université de Ougadougou*, v. 1, no. 7, pp. 69-88.

Kamdem, M., Tounsi, G., & Yodjin, F. (2022). Use of improved fireplaces and vegetation protection in Cameroon. *ESI Preptints*, pp. 688-709. https://doi.org/10.19044/esipreptint.9.2022.p688

Karam, D. B. (2008). *Mechanisms of desert aerosol uplift in West Africa.* PhD thesis in Environmental Sciences, Université Pierre et Marie-Curie - Paris VI, 312 p.

Karen, H.-H. & Thiombiano, A. (2000). Perception of endangered species in a gourmantché environment (eastern Burkina Faso). *Berichte des Sonderforschungsbereichs 268, Band 14, Frankfurt a. M. 2000*, pp. 285-297.

Kiéma, S. (2007). *Extensive livestock farming and biodiversity conservation in the protected areas of western Burkina Faso. A look at their history, current management challenges and vegetation dynamics*. PhD thesis in Physiology and biology of organisms, populations and interactions, Université d'Orléans, 659 p.

Kiéma, S., & Fournier, A. (2007). Use of three protected areas by extensive livestock farming in western Burkina Faso ? In Fournier, A., Sinsin, B., Mensah, G. A. (Eds.). *Quelles aires protégées pour l'Afrique de l'Ouest? Conservation de la biodiversité et développement*, IRD, Collection Colloques et séminaires, Paris, pp. 499-506.

Koffi, K. J. M. (2014). Stylized facts of forest governance in Francophone Sub-Saharan Africa: denial of freedoms and sustainability issues. *Éthique et économique,* UMI Résiliences (IRD-CIRES/France- Côte d'Ivoire) and Université de Bouaké, pp. 145-167.

Koné, M., Aman, A., Adou Yao, C. Y., Coulibaly, L., & N'Guessan, K. E. (2007). Diachronic monitoring by spatial remote sensing of woody cover in a Sudanian savannah environment in Côte d'Ivoire. *Revue télédétection*, v. 7, n° 1-2-3-4, Abingdon-on-Thames, United Kingdom, pp. 433-446.

Korbéogo, Y. S. C. (2018). *Dynamique des ressources forestières et stratégies REDD+ de préservation de la forêt classée de Koulbi (Sud-Ouest du Burkina Faso).* Master's thesis in Geography, Joseph KI-ZERBO University, 97 p.

Koumoi, Z. & Laré, L. Y. (2014). Impact of land tenure on structure and functioning in agrosystems in Tem country. *Rev. Sc. Univ. Lomé (Togo)*, v. 1, no. 11 (special), pp. 85-104.

Ky, K. J. M. (2010). *L'état des ressources végétales pourvoyeuses de produits forestiers non ligneux de la forêt de Bissiga, Centre-Est du Burkina Faso*. PhD thesis in Applied Biological Sciences, University of Ouagadougou, 171 p.

Landmann, G. (1988). How to assess the vitality of a tree or forest stand? *R.F.F.,* v. XL, n° 4, Paris, pp. 265-284.

Laurent, M. C. (2011). *La gestion participative des forêts, une notion devenue multiscalaire et une application différenciée sur le terrain: le cas du Burkina Faso*. Research dissertation, Institut d'Etude Politique de Toulouse, 108 p.

Lebrun, J. (1960). Sur une méthode de délimitation des horizons et étages des végétations des montagnes du Congo oriental. *Bulletin du Jardin Botanique de l'Etat à Bruxelles*, v. 30, Fasc. 1, pp. 75-94.

Lepart, J. (1997). De la diversité spécifique à la biodiversité, les raisons d'un succès. *Definition et diverses perceptions de la biodiversité,* Centre d'écologie fonctionnelle et évolutive-CNRS, Montpellier cedex, pp. 4-10.

Lokossou, R. S., Akouehou, G. S., Avononmadegbé, M. L., & Orou Matilo, A. (2018). Modes of land management in the buffer zone of the Pendjari Biosphere Reserve. *Journal of Applied Biosciences*, v. 124, pp. 12433-12445.

Ly, I. (2001). *Trends in wildlife law and protected areas in West Africa.* FAO Legal Centre online, Rome, 25 p.

Madec, J.-H. (1997). Legislation and jurisprudence: Looking back: French tropical forest legislation. *Rev. For. Fr. XLIX*- 1-1997, Paris, pp. 69-78.

Madjimbe, G., Goalbaye, T., Belem, M. O., & Ngarikla, B. (2018). Assessment of woody resources and their exploitation as fuelwood and service wood in the Department of Barh-Kôh in southern Chad.

International Journal of Biological and Chemical Sciences, v. 12, no. 6, University of Dschang, Cameroon, pp. 2856-2870.

Madon, G. (2017). Wood, energy of first necessity in Africa: a resource too often neglected. *AFRIQUE CONTEMPORAINE*, n° 261-262, pp. 201-222.

Mahamane, A., Mahamane, S., & Lejoly, J. (2007). Phenology of some woody species in the "W" National Park of Niger. *Sécheresse,* v. 1E, n° 4, Paris, pp. 1-13.

Mahamane, A. (2006). Floristic, phytosociological and phythogeographic studies of the vegetation of the W Regional Park of Niger. *Acta Botanica Gallicia*, v. 153, no. 2, pp. 265-269.

Mainguet, M. (1995). *L'homme et la sécheresse.* Masson, Paris, France, 335 p .

Marcon, E. (2018). *Biodiversity measurements.* Master, Kourou, France 284 p.

Mayaux, P., Eva, H., Fournier, A., Sawadogo, L., Palumbo, I., & Grégoire, J.-M. (2003). Apport des techniques spatiales pour la gestion des aires protégées en Afrique de l'Ouest. *Séminaire régional sur l'aménagement et la gestion des aires protégées d'Afrique de l'Ouest,* Parakou, Bénin, April 14-19, 2003, 8 p.

Mbayngone, E., Thiombiano, A., Hahn-Hadjali, K., & Guinko, S. (2008). Woody structure of plant formations in the Pama Reserve (southeastern Burkina Faso, West Africa). *Flora et vegetatio Sudano-Sambesica,* v. 11, Frankfurt, pp. 25-34.

Mbayngone, E., & Thiombiano, A. (2011). Degradation of protected areas by plant resource exploitation: the case of the Pama partial wildlife reserve, Burkina Faso (West Africa). *Fruits,* v. 66, Les Ulis, France, pp. 187-202.

MEEVCC (2019). *Drivers of deforestation and forest degradation in Burkina Faso. Volume 1: the current trend.* Study report, Ouagadougou, Burkina Faso, 177 p.

MECV (2007a). *Situation des forêts classées du Burkina Faso et plan de réhabilitation.* Ouagadougou, 48 p.

MECV (2007b). *Programme d'action national d'adaptation à la variabilité et aux changements climatiques, PANA du Burkina Faso.* CONEDD, Ouagadougou, 76 p .

MECV (2009). *Plan d'aménagement et de gestion des réserves partielle et totale de la faune de Bontioli et du corridor Bontioli-Koulbi.* PROGEREF, Ouagadougou, 122 p.

MECV (2010a). *Stratégie nationale de valorisation et de promotion des produits forestiers non ligneux.* Final version, Ouagadougou, Burkina Faso, 72 p.

MECV (2010b). *Economic valuation of the environment and natural resources in Burkina Faso: economic-environmental analysis at national level (phase 1).* Final report, Ouagadougou, Burkina Faso, 78 p.

MEDD (2011). *Forest Investment Program (PIF- Burkina Faso).* Ouagadougou, 70 p.

MEDD (2012). *Strategy and action plan of the great green wall initiative in Burkina Faso.* Ouagadougou, 117 p.

MEDD (2013). *National wetland policy of Burkina Faso.* Ouagadougou, Burkina Faso, 110 p.

MEE (1996). *Forum sur les politiques forestières en Afrique sub-saharienne: communication du Burkina Faso*. Ouagadougou, 13 p .

MEE (1998). *Politique forestière nationale*. Ouagadougou, Burkina Faso, 55 p.

MEEVCC (2017). *Avant-projet d'aménagement et de gestion actualisé des réserves partielle et totale de faune de Bontioli.* Final report, Ouagadougou, 88 p.

Menaut, J. C., Lepage, M., & Abbadie, L. (1995). Savannas, woodlands and dry forests in Africa. *S.H. Bullock, H.A. Mooney and E.E. Medina (Editors), Seasonally dry tropical forests,* Paris, pp. 64-92.

Mengue-Medou, C. (2002). Protected areas in Africa: perspectives for their conservation. *Vertigo- la revue électronique en science de l'environnment*, v. 3, n° 1, https://doi.org/10.4000/vertigo.4126

Montagne, P., Nouvellet, Y., & Bertrand, A. (2005). Nouvelles politiques forestières, interventions multiples de nouveaux acteurs privés et renouvellement des fonctions des agents des administrations. *L'Etat de la gestion locale durable des forêts en Afrique francophone et à Madagascar,* L'Harmattan, Paris, pp. 96-128.

Montouroy, Y. (2013). Analyzing global forest governance: capturing international change using political sociology. *Les cahiers CRGRNT*, n° 1305, Institut d'études politiques de Bordeaux, France, 15 p.

Morin, X. (2006). *Biogeography of European and North American tree species: determinism and evolution under the effect of climate change.*

PhD thesis in Population Biology and Ecology, Université de Montpellier, Sciences et Techniques du Languedoc, 468 p.

Moumouni, Y. I., Arouna, O., & Zakari, S. (2017). Floristic diversity and structure of plant formations in the Borgou-Nord phytogeographic district of Benin (Bagou arrondissement sector). *Homme et société,* n° 7, Paris, pp. 63-80.

Muoghalu, I. J. (2014). Vulnerability of biophysical and socioeconomic systems of West and Central African savannas and woodlands to climate change. *African Forest Forum*. WorkingPaperSeries, v. 2, no. 14, Nairobi, Kenya, 35 p.

United Nations. (1992). Convention on Biological Diversity, New York, 32 p.

Ndambo, M. N. (2014). *Local community participation and sustainable forest management: the case of the Luki Biosphere Reserve in the Democratic Republic of Congo.* PhD thesis in Forest Sciences, Université Laval, Québec, Canada and Université de Kinshasa, République Démocratique du Congo, 205 p.

Ngo, M. R., Sanou, P., Touré, I., Tchindjang, M., & Makak, J. S. (2018). Diachronic analysis of land use for the design of a geo-referenced database for monitoring territorial dynamics in the rural commune of Koumbia, Burkina Faso. *Revue Scientifique et Technique Forêt et Environnement du Bassin du Congo*, v. 10, pp. 23-35.

Nouazeur, Z. (2020). The resumption of rains and the recrudescence of floods in Sahelian West Africa. *Physico-Géo*, v. 15, n°1, pp. 89-109.

Oloukoi, J., Yabi, I., & Johnson, D. (2014). Influence of climatic and topographic factors on wildfire risk in Central Benin. *Journal of Geospatial Science and Technology*, v. 1, no. 1, pp. 64-86.

WHO (2002). *WHO strategies for traditional medicine 2002-2005,* WHO, Geneva, 78 p.

Orsenna, E., & Petit, M. (2011). *Climat : une planète et des hommes,* Le cherche midi, Paris, France, 333 p.

Ouattara, D., Kouamé, D., Tiébré, M.-S., Cissé, A., & N'Guessan, K. E. (2016). Floristic diversity and plant uses in the Sudanian zone of northwest Côte d'Ivoire. *Journal of Animal and Plant Sciences,* v. 31, no. 1, Nairobi, Kenya, pp. 4815-4830.

Ouédraogo, A., Thiombiano, A., Hahn-Hadjali, K., & Guinko, S. (2006). Juvenile stand structure and woody regeneration potential in eastern Burkina Faso. Etude sur la flore et la végétation du Burkina Faso et des pays avoisinants. 10, pp. 17-24.

Ouédraogo, B. (2015). *Adaptation strategies of agropastoralists to climate variability in the Yakouta watershed (Burkina Faso)*. PhD thesis in Geography, University of Ouagadougou, 257 p.

Ouédraogo, F. C. (2006). *Food vulnerability in Burkina Faso.* L'Harmattan, Paris, 226 p .

Ouédraogo, K. (2001). *L'étude prospective du secteur forestier en Afrique (FOSA) : Burkina Faso,* Rome, 39 p.

Ouédraogo, K. (2004). *Regional workshop to capitalize on Sahelian experience in the management of natural forests for wood production: Burkina Faso's contribution.* Programme Régional de Promotion des

Énergies Domestiques et Alternatives au Sahel (PREDAS), Ouagadougou, 32 p.

Ouédraogo, L. (2012). *Gestion de l'eau et adaptation des populations au changement climatique dans le bassin versant de Yakouta (Sahel du Burkina Faso)*. PhD thesis in Geography, Université Abdou Moumouni, Niger, 245 p.

Ouédraogo, M. (2011). *Diffusion et appropriation des textes juridiques dans le cadre de la gouvernance locale des ressources forestières: cas de la zone villageoise d'intérêt cynégétique (ZOVIC) de Kompienbiga.* Mémoire d'ingénieur en Vulgarisation agricole, IDR, Université Polytechnique de Bobo-Dioulasso, 111 p.

Ouédraogo, R., Soara, A. E., Oueda, A. (2015). Description du peuplement piscicole du lac sahélien de Higa, un site Ramsar du Burkina Faso, Afrique de l'Ouest. *Journal of Applied Biosciences,* v. 95, Nairobi, Kenya, pp. 8958-8965.

Ouédraogo, S. B. (2011). *Forests in Burkina Faso.* Atelier sur le dialogue forestier du 12 au 15 septembre 2011, Ouagadougou/Burkina Faso, Power Point, 23 p.

Ouédraogo, T. L. (2015). *Resist, adapt or disappear? Les paysanneries face aux mutations agricoles et foncières: une analyse à partir des provinces du Ziro et de la Sissili au sud du Burkina Faso.* Doctoral thesis in Geography, Université Paris Panthéon Sorbonne, 446 p.

Ouédraogo, M., Ouédraogo, D., Thiombiano, T., Hien, M., & Likke, A. M. (2013). Economic dependence on non-timber forest products: the case of households living along the Boulon and Koflandé forests, southwest Burkina Faso. *Journal of Agriculture and Environment for International Development-JAEID*, v. 107, no. 1, pp. 45-72.

Ouharon, A. (2006). Population and environment: uncertain links. *Nature-Sciences-Sociétés,* v. 14, pp. 168-173.

OSS (2015). *Burkina Faso : atlas d'occupation du sol - Projet d'Amélioration de la résilience des populations sahéliennes aux mutations environnemental-REPSAHEL*, Tunis Carthage, Tunisia, 138 p.

Ouoba, A. P. (2013). *Climate change, vegetation dynamics and peasant perception in the Burkinabe Sahel.* Unique doctoral thesis in Geography, University of Ouagadougou, 305 p.

Ouoba, P. (2007). *La protection de la végétation: cas de la forêt classée de Nahouri*. Master's thesis in Geography, University of Ouagadougou, 109 p.

Ouoba, P. A., Da, D. E. C., & Paré, S. (2014). Effects of climate variability and anthropogenic activities on land-use dynamics at Gandafabou in the Burkinabe Sahel. *Cahiers de CBRST,* n° 6, pp. 68-94.

Ouoba, P. A., Da, D. E. C., & Paré, S. (2014). Local perception of woody stand dynamics over the last twenty years in the Burkinabe Sahel. *Vertigo- la revue électronique des sciences de l'environnement,* v.14, n° 2, 13 p. http://vertigo.revues.org

Ozer, A., & Ozer, P. (2005). Desertification in the Sahel: climatic or anthropogenic crisis? *Bull. Séanc. Acad. R. Sci. Outre-Mmer Meded. Zitt. K. Acad. Overzeese Wet.* v. 51, no. 4, pp. 395-423.

Ozer, P. (2006). Desertification, a false problem? *Vox geographica*, *Les Cafés Géographiques*, Université de Liège, Brussels, 8 p.

Ozer, P., Hountondji, Y.-C., Niana, A. J., Karimoune, S., Laminou Manzo, O., & Salmon, M. (2010). Desertification in the Sahel: history and perspectives. *Bulletin de la Société Géographique de Liège*, v. 54, pp. 69-84.

Palé, O. K. F. (2015). Agricultural colonization and land issues in the Bougouriba valley (Burkina Faso). *Revue des questions foncières,* v. 1, n° 15, pp. 45-65.

Palé, S. (2020). *Morphohydrology of the Poni watershed (Southwest Burkina Faso).* Doctoral thesis in Geography, Joseph KI-ZERBO University, 275 p.

Pallo, F. J. P., & Sawadogo, N. (2010). An attempt to correlate climate, vegetation and soil characteristics in Burkina Faso. *International Journal of Biological and Chemical Sciences,* v. 4, no. 5, University of Dschang, Cameroon, pp. 1839-1850.

Papy, F., & Torre, A. (2003). Quelles organisations territoriales pour concilier production agricole et gestion des ressources naturelles *(From agronomy and economics questions). Études et Recherches sur les systèmes agraires et le développement,* n° 33, pp. 151-169.

Pascal, J.-P. (2003). Description et dynamique de milieux forestiers: notions sur les structures et dynamiques des forêts tropicales humides. *Rev. For. LV-* special issue 2003, Paris, pp. 118-130.

Poan, D. E. (2013). *Documentation and interpretation of inter-seasonal variability in the African monsoon; application to forecasting.* PhD thesis, University of Toulouse, 248 p.

Poda, E. N. (1998). Le sacré et les lieux sacrés: voie privilégiée de sauvegarde de l'environnement. Cases of Tio and Nagarpoulou. *Aménagement intégré des forêts naturelles des zones tropicales sèches de l'Afrique de l'Ouest.* Proceedings of the International Seminar, November 16-20, 1998, Ouagadougou, Burkina Faso, pp. 263-265.

Pratt, W. K. (1991). *Digital image processing.* 2(nd) ed., Willey-interscience, 698 p.

PREDAS. (2005). *Guide méthodologique d'aménagement forestier villageois pour la production du bois énergie.* Collection: Les Guides Techniques du PREDAS. CILSS, Ouagadougou, 66 p.

Puyo, J.-Y. (2001). Sur le mythe colonial de l'inépuisabilité des ressources forestières (Afrique occidentale française / Afrique équatoriale française, 1900-1940). *Cahier de Géographie du Québec,* v. 45, no. 126, pp. 479-496.

Raffin, J.-P. (2005). From nature protection to biodiversity governance. *Ecologie et politique,* n° 30, Paris, pp. 97-109.

Reboul, C. (1976). Mode de production et système de culture et d'élevage. *Économie Rurale.* 112. March-April 1976, Paris, pp. 55-65.

Ribot, J.-C. (2001). History of forest management in West Africa. Or: how "science" excludes farmers. *Les dossiers d'*Haramata, IIDE publications, Paris, France, 16 p.

Rondeux, J. (1993). *Measuring trees and forest stands.* Les Presses Agronomiques de Gembloux, Belgium, 512 p.

Rondeux, J. (2021). *Measuring trees and forest stands.* Les Presses Agronomiques de Gembloux, Belgium, 738 p.

Rouamba, T. (1997). *Essai de détermination de tailles optimales de plots dans les inventaires forestiers : " cas de la forêt classée de Tuy ".* Mémoire d'Ingénieur de Développement Rural, Université de Ouagadougou, Centre Universitaire polytechnique de Bobo-Dioulasso, 90 p.

Sangli, G., & Tallet, B. (2012). Demographic dynamics and settlement distribution. Vers une nouvelle géographie du peuplement dans le Grand Sud-Ouest du Burkina Faso? *Espace Populations sociétés*, [online], 1, pp. 99-110.
http://eps/revues.org/4896

Sanguin, A.-L., & Pĕna, O. (1986). *Concepts et méthodes de la Géographie*. Guérin Éditeur, Montreal, Canada, 206 p.

Sanou, L., Koala, J., Ouédraogo, S., & Ouattara, B. (2022). Perceptions, ecosystem services and vulnerability of multiple-use woody species of the 20th Ramsar site in Burkina Faso, West Africa. *Afrique SCIENCE*, v. 20, no. 3, pp. 25-40.

Sarr, M. A. (2008). Rainfall variability in West Africa: vegetation dynamics from satellite images. Example of the Ferlo watershed (Senegal). *Climat et Société: Climate and vegetation.* Journées de Climatologie -Nantes, March 13-14, 2008, pp. 57-76.

Savadogo, O. M., Ouattara, K., Barron, J., Ouédraogo, I., Gordon, L., Enfors, E., & Zombré, N. P. (2015). States of Sahelian ecosystems: revegetation, loss of diversity and soil quality. *Afrique SCIENCE*, v. 11, no. 5, Abidjan, Côte d'Ivoire, pp. 433-446.

Savadogo, S., Kaboré, A., & Thiombiano, A. (2017). Plant characterization, typology and functions of sacred woods in Burkina Faso. *International Journal of Biological and Chemical Sciences*, v. 11, no. 4, University of Dschang, Cameroon, pp. 1497-1511.

Savadogo, S. (2018). *Impacts of climate change and variability and anthropogenic factors on the availability of non-timber forest products in the Sudano-Sahelian domain of Burkina Faso: the cases of Kokologho and Ténado*. Doctoral thesis in Geography, Université Ouaga I Professeur Joseph KI-ZERBO, 404 p.

Savadogo, S., Ouoba, P. A., & Da, D. E. C. (2019). NTFP extraction and threat to provider species in Burkina Faso: cases of Kokologho and Ténado in the Centre-Ouest region. *Rev. Ivoir. Sci. Technol.* v. 33, Abidjan, Côte d'Ivoire, pp. 240-258.

Savadogo, S., Sambéré, O., Ouédraogo, I., & Thiombiano, A. (2020). Exotic indicator species of sacred wood degradation in Burkina Faso. *Geo-Eco-Trop,* v. 44, no. 3, Liège, Belgium, pp. 409-426.

Sawadogo, L. (2006). *Adapting sustainable forest management approaches to social, economic and technological aptitudes in Africa. Le cas du Burkina Faso.* Ouagadougou, 59 p.

Sawadogo, L. (2007). *Status of biodiversity and timber production in the NAZINON forest management site after twenty years of management practices.* Center for International Forestry Research, Ouagadougou, Burkina Faso, 52 p.

Sawadogo, L. (2009). *Influence des facteurs anthropiques sur la dynamique de la végétation des forêts classées de Laba et de Tiogo en zone soudanienne du Burkina Faso.* Thèse de doctorat 3ème Cycle ès Sciences Naturelles, Université de Ouagadougou, 181 p.

Sebilotte, M. (1974). Agronomie et agriculture: essai d'analyse des tâches de l'agronome. *Cah. ORSTOM sér. Biol.* n° 24, Paris, France, pp. 3-25.

SCR. (2013). *Ramsar Convention Manual: Guide to the Convention on Wetlands (Ramsar, Iran, 1971)*. 6ᵉ edition. Gland, Switzerland, 120 p.

Semini, T. G. (2007). *Integrating gender issues into the forestry sector in Africa: Burkina Faso*. Food and Agriculture Organization of the United Nations, Rome, 66 p.

Somda, S., & Somé, L. (2009). *Monographie de la région du Sud-Ouest* RGPH-2006, INSD, Ouagadougou, 152 p.

Somé, B. B. (2015). *Stratégies d'adaptation des populations riveraines des forêts galeries du Mouhoun aux changements climatiques au Burkina Faso: cas des femmes Dagara de Dissin dans la province du Ioba.* Master's thesis in Sociology, Université Ouaga I Professeur Joseph KI-ZERBO, 106 p.

Somé, E. S. (2013). *Impact des activités anthropiques sur la structure végétale des réserves de Bontioli (Sud-Ouest du Burkina Faso)*. DEA dissertation, IDR, Université Polytechnique de Bobo-Dioulasso, 34 p.+annexes.

Somé, T. N. (2012). *Gestion participative et sécurité alimentaire dans le Chantier d'Aménagement Forestier de Bougnounou.* Master's thesis in Geography, University of Ouagadougou, 93 p. +annexes.

Somé, T. N. (2016). *Effets des actions anthropiques et des changements climatiques sur la dynamique des forêts naturelles sèches dans la région Centre-ouest du Burkina Faso*. Master's thesis in Geography, Université Ouaga I Pr Joseph KI-ZERBO, 84 p.

Somé, Y. S. C. (2010). *Modeling the spatial distribution of molecular forms M and S of Anopeles gambiae sensu stricto with GIS and spatial analysis*. Doctoral thesis in Geography, Université d'Orléans, 209 p.

Somé Y. S. C., Akaffou, Y. F. W., & Kaguembega-Müller, F. (2011). Evaluation of biodiversity restoration in fragile ecosystems: the case of newTree Burkina set-asides. *Actes des Collectes de Yaoundé*, pp. 1-14.

Secretariat of the Convention on Biological Diversity (2004). *Programme of Work on Protected Areas (CBD Programme of Work)*. Secretariat of the Convention on Biological Diversity, 34 p.

SP/CONAGESE. (1999). *Burkina Faso's national biodiversity strategy and action plan*. Ouagadougou, Burkina Faso, 126 p.

Solly, B., Dieye, E. H. B., Mballo, I., Sy, O., Sane, T., & Thior, M. (2020). Spatiotemporal dynamics of forest landscapes in southern Senegal: the case of the Vilingara department. *Physio-Géo*, v. 15, pp. 41-67. https://doi.org/10.4000/physio-geo.10634

Souley, R. A., Diadié, H. O., Abdelkader, A. S., Yacoubou, B., & Balla, A. (2020). Biochemical and microbial characterization of néré (*Parkia biglobosa*) and Guinea sorrel (*Hibiscus sabdariffa*) soumbala produced in Niger. *European Scientific Journal,* v. 16, n° 3, pp. 224-243.

Tankoano, B. (2012). *Suivi diachronique de la couverture ligneuse dans la forêt classée de Koulbi à l'aide de la télédétection et des systèmes d'information géographique (SIG).* Mémoire d'Ingénieur du Développement Rural (IDR), Université Polytechnique de Bobo-Dioulasso, 84 p.+annexes.

Tankoano, B., Sanon, Z., Hien, M., Dibi, N. H., Yaméogo, J. T., & Somda, I. (2016). Anthropogenic pressure and plant dynamics in the Tiogo classified forest in Burkina Faso: contribution of remote sensing, *TROPICULTURA,* v. 34, no. 2, Liège, Belgium, pp. 193-207.

Tapsoba, A. *Valorisation économique des produits forestiers non ligneux au Burkina Faso : cas de Parkia biglobosa (néré).* Master's thesis in Innovation and Development in Rural Areas, Université de Ouagadougou-CNRST, 66 p. www.agrinovia.org

Tapsoba, B. (2016). *Land occupation of forestry developments and surroundings in Ziro province from 1992 to 2013 and beyond.* Master's thesis in Geography, University of Ouagadougou, 96 p.

Tianhoun, V. F. (2012). *Conservation of biological diversity and types of use of the Toroba classified forest and its zone of influence*. Mémoire d'Ingénieur de Développement Rural, Université Polytechnique de Bobo-Dioulasso, 80 p.

Tiesse, C. B., Wandan, E. N., & N'da, H. D. (2017). Contribution of remote sensing for spatio-temporal monitoring of land use in the Tonkpi mountain region (Côte d'Ivoire). *European Scientific Journal*, v. 13, no. 15, pp. 310-329.

Tinyar, S. (2003). *Connaissance des pluies au Burkina Faso: variabilités temporelles et épisodes secs en saison des pluies.* Master's thesis in Geography, University of Ouagadougou, 128 p.+ appendices.

Thiombiano, A., Glèglé Kakaï, R., Bayen, P., Bousim, J. I., & Mahamane, A. (2016). Forest inventory methods and devices in West Africa: state of the art and proposal for harmonization. *Annale des Sciences Agronomiques,* 20-special, Projet Undesert-UE, Cotonou, Benin, pp. 15-31.

Thiombiano, A., & Kampmann, D. (2010). *Atlas de la Biodiversité de l'Afrique de l'Ouest,* VOLUME II : Burkina Faso, Ouagadougou and Frankfurt/Main, 592 p.

Traoré, D. (2021). *Mutation d'usage des terres et dynamique de la biodiversité végétale ligneuse dans la commune de Kangala, Burkina Faso*. Unique PhD thesis in Geography, Joseph KI-ZERBO University, 270 p.

Tristan, D., Cassedane, X., Even, M.-A., & Vert, J. (2009). *Farming systems and technical itineraries.* Fiche-variable. CENTRE D'ETUDE ET DE PROSPECTIVE, Paris, France, 9 p.

Tubiana, J. (2019). *Protected areas, conflict and insecurity: understanding the situation and defining the rules*. Papaco, 24 p.

IUCN (1994). *Guidelines for Protected Area Management Categories*. Gland, Switzerland, Cambridge, IUCN (CNAPPA)/WCMC, 261 p.

UICN (1994). *Parks for biodiversity: policy guidance based on experience in APC countries*. Brussels and IUCN, Gland Switzerland and Cambridge, U.K.

UICN/PACO (2009). *Evaluation of protected area management effectiveness: protected areas in Burkina Faso*. IUCN, Gland, Switzerland and Cambridge, UK, 84 p.

UICN/PACO (2009). *Les aires communautaires en Afrique de l'Ouest: quelle contribution à la concertation?* IUCN, Gland, Switzerland and Cambridge, UK, 68 p.

UICN/PACO (2010). *Legal and institutional assessment for the establishment of conditions to improve the management of protected areas in West Africa.* Ouagadougou, Burkina Faso: IUCN/PACO, 93 p.

UICN/PAPACO (2012). *Strengthening conservation in Africa's protected areas. Synthesis of the Weotenga meeting (Burkina Faso-25 to 27 October 2011)*. Ouagadougou, Burkina Faso, 58 p.

Valéa, F. (2010). *Study of fires in Burkina Faso. Multiscale approach to active fires and burned areas.* Doctoral thesis in Geography, Université Caen-Basse Normandie, 412 p.

Valéa, F., & Ballouche, A. (2012). Bushfires in West Africa: environmental constraints or environmental management tools? The example of Burkina Faso. *Territoires d'Afrique*, vol. 3, Dakar, Senegal, pp. 36-47.

Velle, M. G. P. V., & Zandée, R. (2002). Highlighting vicariant methodologies in vicariance biogeography. *Biosystema 20, Systématique et biogéographie*, pp. 7-14.

Vischel, T., Lebel, T., Panthou G., Rossi, A., & Martinet, M. (2015). The return of a wet period in the Sahel? Observations and perspectives. In Sultan, B., Lalou, R., Amadou, S. M., Oumarou, A., & Soumaré, M. A. (eds.). *Les sociétés rurales face aux changements climatiques et environnementaux en Afrique de l'Ouest*. IRD, Marseille, pp. 43-60.

Vissin, E. W. (2007). *Impact of climatic variability and surface state dynamics on runoff in the Benin basin of the Niger River*. PhD thesis in Hydroclimatology, Université de Bourgogne, 311 p.

Wala, K. (2010). The vegetation of the Atakora range in Benin: floristic diversity, phytosociology and human impact. Summary of thesis. *Acta Bot. Gallica,* v. 157, no. 4, pp. 793-796.

Wallace, G. N. (1998). *Toward a principled evaluation of ecotourism ventures.* Colorado Strate University, Denver.

Woro, N. O., & Sinsin, B. (2007). Ecotourism in forest areas: the case of the Lama forest (Benin). In Fournier, A., Sinsin, B., Mensah, G. A. (Eds.). *Quelles aires protégées pour l'Afrique de l'Ouest? Conservation de la biodiversité et développement*, IRD, Collection Colloques et séminaires, Paris, pp. 449-478.

Yamba, B. (2004). Les mutations des systèmes agraires et des modes d'usage des ressources naturelles dans la zone centrale du Niger. *Revue de Géographie Alpine*, tome 92, n° 1, Grenoble, pp. 97-110 .

Yaméogo, A., Somé, Y. S. C., Sirima, A. B., & Da, D. E. C. (2020). Land use and soil erosion in the upper Sissili watershed, Burkina Faso. *Afrique SCIENCE,* v. 17, no. 5, pp. 43-56.

Yaméogo, T. J., Somé, A. N., & Hien, M. (2009). Preliminary study for the restoration of degraded soils in the Sudanian zone of Burkina Faso. *Sécheresse,* v. 20, no. 1, pp. 32-38.

Yanogo, P. I. (2013). *Les stratégies d'adaptations aux aléas climatiques autours du Lac Bagré (Burkina Faso)*. Thèse de doctorat unique de Géographie, Université d'Abomey-Calavi, 302 p.

Yaovi, R. C. (2017). *Floristic diversity and ecosystem services of the Kou class forest in southwest Burkina Faso*. Master's thesis in Natural Resource Management, Université Nazi BONI, Bobo-Dioulasso, 88 p.

Yaro, K. (2013). *The induced effects of mining on natural resources.* Master's thesis in Geography, University of Ouagadougou, 64 p.

Yonkeu, S. (2006). *Strategic environmental assessment of Burkina Faso's food security, nutrition and soil fertility program.* Institut de l'environnement et de l'énergie de la Francophonie, Groupe EIER-ETSHER, Burkina Faso, Ouagadougou, 12 p.

Yonkeu, S., Da, D. E. C., & Soule, M. (1995). *Activités agro-pastorales et dégradation du couvert végétal dans le Centre-Nord du Burkina Faso : cas du bassin versant de Bourzanga,* EIER-ETSHER, Université de Ouagadougou, POWER POINT, 15 p.

Zida, D., & Ouadba, J.-M. (1998). Dynamics of natural woody vegetation in a managed forest area. Cas de la forêt de Maro. *Aménagement intégré des forêts naturelles des zones tropicales sèches de l'Afrique de l'Ouest.* Proceedings of the International Seminar, November 16-20, 1998, Ouagadougou, Burkina Faso, pp. 209-218.

Zonou, B. (2008). *Interactions between land tenure issues and socio-territorial identities in western Burkina Faso.* Doctoral thesis in Rural Development, University of Toulouse, 350 p.

Zorom, A., Ouoba, P., A., Somé, W., & Somé, Y. S. C. (2022). Relationships between rainfall and vegetation production during period 1999-2019 in Burkina Faso. *International Journal of Progressive Sciences and Technologies (IJPSAT)*, v. 35, no. 1, pp. 305-314.

Zougouri, S. (2006). Tutor-migrant relationship and land practices: the migrants caught between the power of the land and the power of the gods in Bougnounou (Ziro province, Burkina Faso). *Colloque International, Les frontières de la question foncière- At the frontier of land issues*, Montpellier, France, 17 p.

Zoungrana, T. P. (1995). Sécheresse et dynamique des agrosystèmes dans la plaine centrale du Burkina/ Drought and dynamics of agrosytems in the central plain of Burkina. *Revue de géographie de Lyon,* v. 70, no. 3-4, pp. 247-254.

http://fr.wikipedia.org/wiki/Accord_de_Paris_sur_le_climat
http://fr.wikiversity.org
www.biodiversite2010.ch/comprendre/definition/
www.cnle.gouv.fr/Definition-de-la-pauvrete.html
www.logement.gouv.fr/le-reseau-national-des-amenageurs
http://environnement.doctissimo.fr/faune-et-flore/
http://fr.wikipedia.org/wiki/Ressource_naturelle
https://fr.wikipedia.org/wiki/Biogéographie
https://fr.wikipedia.org/Liste_des_formations_végétales
https://www.aquaportail.com/definition-9830-formation-vegetale.html

The Constitution of June 11, 1991 of Burkina Faso.
LOI N°2001-013/AN creating the regions.
LOI N° 003-2011/AN portant Code forestier au Burkina Faso, Ouagadougou, 53 p.
LOI N°034-2009/AN portant Régime foncier rural, Ouagadougou, Burkina Faso, 31 p.
LOI N° 063-2015/CNT portant satut du personnel du cadre paramilitaire des Eaux et Forêts, Ouagadougou, Burkina Faso, 61 p. + appendix.

APPENDICES

Appendix 1: Survey questionnaire for heads of households in study sites

Order no.	Questions	
	I. Identification of the head of household	
1	Gender	Male □ ; Female □
2	Age :	To be specified
3	Marital status	Married □; divorced □; widowed □; single □
4	Number of children	To be specified
5	Profession:	To be specified
6	Level of education	None □; literate □; primary □; secondary □; higher □
	II. Agricultural and pastoral activities and their impact on the reserves	
7	What is your main activity	Agriculture □; livestock □; forestry □; hunting □; fishing □; other □ please specify
8	What are your secondary activities?	Agriculture □; livestock □; forestry □; hunting □; fishing; other □ please specify
9	What speculations do you produce?	-Cereals : Millet □; white sorghum □; red sorghum □; corn □ -legumes: beans □; cowpeas □ -Income speculation Cotton □ ; peanuts □ ; sesame □ ; Market garden speculation Tomato □; onion □; bell pepper □; lettuce □; squash □; sorrel □; other □ please specify
10	What means of production do you use?	Tractor □; plow □; daba □; hoe □; pickaxe □; cutter □ ; other □ please specify
11	What the average size of your field?	≤ 1ha □; [1ha-3ha] □ ; [4ha- 6ha] □ ; [7ha-10ha] □ ;[11ha-15ha] □; [16ha-20ha] □ ; >20ha
12	How did you acquire it?	Loan □; gift □; inheritance □; purchase □; other □ please specify
13	Is there still enough bush to open up new fields?	Yes □; no □; abstention □

14	If so, which areas are subject to clearing?	Bush □; old fallow □; forest □
15	What techniques do you use to open up a new field?	Selective felling of trees and shrubs followed by ploughing □; systematic felling of trees and shrubs followed by ploughing □; ploughing without felling of trees and shrubs □; other techniques □ to be specified
16	If not, what are you doing to fertilize your soils, which are already subject to intense agricultural activity?	Use of organic manure □; use of chemical fertilizers □; zaï □; living hedge □; stone cordons □ ; other □ please specify
17	Do you fallow?	Yes □; no □; abstention □
18	If so, how often?	[1-3y] □ ; [3-5y] □ , 5-10y □ ; 10-15y □ ; 15y □ ; over 15y □
19	If not, why not?	Insufficient land □ ; reduction of agricultural space □ ; other reasons □ to be specified
20	Is cereal production sufficient all year round?	Yes □; no □; abstention □
21	If not, what are you doing to make up the deficit?	Purchase of food □ ; use of NTFPs to supplement diet □ ; other to be specified
22	What type of breeding do you do?	Family (hutched) livestock □ ; transhumant livestock □ ; fattening □
23	What is the composition of your livestock?	Cattle □; sheep □; goats □; asinine species □, camels □; pigs □; poultry □; others □ to be specified
24	What types of feed do you use to feed your animals?	SPAI □; cultivated forage □; wild forage □; residues of agricultural products □; other □; other to be specified
25	What are your grazing areas?	Former fallow land □; total reserve □; partial reserve □; off-reserve bush □; other areas □ to be specified
26	Do you prune trees to feed your livestock?	Yes □; no □; abstention □
27	If so, which plant species are involved?	A quote from

28	What are the cutting zones?	Forest □; fields □; old fallow land □; plantations □
III. Socio-economic profitability and cultural promotion of reserves		
29	Why do you use forest products?	Food □; trade □; health □; housing construction □; livestock feed □; other □ please specify
30	What forest products do you consume (specify parts)?	To be specified
31	Name the forest products used in your health care (specify the parts)?	To be specified
32	Which plant and herb species are used in the construction of your homes?	To be specified
33	What type of energy do you use for cooking?	Wood □; charcoal □; butane gas □; biogas □; harvest residues □; other □ please specify
34	Where does the wood used for cooking in your household come from?	Forest □; old fallow □; field □; plantation □; other □ please specify
35	Do you sell wood?	Yes □; no □; abstention □
36	If so, what type of wood do you sell?	Green wood □; dead wood □
37	Where does it come from?	Forest □; field □; old fallow □; plantation □; bush outside forest
38	If the wood you sell comes from the forest, are you authorized to do so?	Yes □; no□; abstention □
39	What is your average annual income from wood sales?	To be specified
40	Do you market NTFPs?	Yes □; no□; abstention □
41	If yes, name the main NTFPs you market	To be specified
42	What is the average annual income from the marketing of forest products in general?	To be specified
43	What do you do with the money from the forest products trade?	Food□; medical care □; children's schooling □; purchase of clothing and kitchen utensils □; other uses □ please specify
44	What contribution does forestry income make to annual family income?	]0 ; 25%] ;] 25% ; 50%]]50% ; 75%] ;]75% ; 100%]

45	How well do you use forest products to meet your needs?	Very good □; good □; fair □; poor □; very poor □
46	What solutions are you proposing to enhance the economic value of the reserves for the benefit of local communities?	To be specified
47	Are you aware of any religious rituals in the forest?	Yes □; no □; abstention □
48	If so, what time of year do these rites take place?	To be specified
49	Which social strata are concerned?	Customary □; men □; women □; initiates □; children □; others □ to be specified
50	What does the forest mean to you in terms of your traditions?	To be specified
51	If not, have the rites been banned in the forest, or have they never existed in the forest to your knowledge?	Forbidden □ ; never existed □
52	Do you welcome tourists to visit the reserves?	Yes □; no □; abstention □
53	If not, why not?	Undeveloped reserves □; lack of visitor facilities □; lack of advertising on reserves' cultural and tourist potential □; other reasons □ to be specified
54	What solutions do you propose to reverse this situation?	To be specified
55	If so, when do tourists visit the reserves?	To be specified
56	Are there professional guides to direct tourists?	Yes □ ; no □ ; ; abstention □
57	Are there any tourist facilities in the area?	Yes □; no □; abstention □
58	How does tourism contribute to the development of your locality?	To be specified
59	What solutions do you recommend to enhance the cultural value of the reserves for the benefit of local communities?	To be specified

IV. Knowledge of forestry legislation		
60	Do you know that Burkina Faso has forestry legislation?	Yes □ ; no □ ; ; abstention □
61	If not, why not?	Lack of communication from the department in charge of the environment □; illiteracy □; disinterest □; other reasons □ please specify
62	If so, what do you know about this legislation?	To be specified
63	What activities in the forest require authorization from the department in charge of the environment?	Hunting □; greenwood cutting □; fishing □; other □ to be specified
64	What activities are prohibited in the forest?	orpaillage □ ; pêche □ ; élevage □ ; agriculture □ ; feu tardif □ ; autres □ à préciser
66	Are there any plant or animal species whose exploitation is prohibited by law?	Yes □, no □; abstention □
66	If so, which ones?	To be specified
V. Farmers' organizations and reserve resource management		
67	Is there a forest management structure in your village?	Yes □, no □; abstention □
68	If so, what activities does organize?	Awareness raising □; Layon opening □, early fire setting □; forest monitoring □; reforestation □; other activities □ to be specified
69	Are you involved in forest management?	Yes □; no □; abstention □
70	If so, what do you do concretely?	Participation in layon opening □; participation in early fire setting □, participation in forest surveillance □; participation in reforestation □; participation in meetings □, participation in other activity □ to be specified

71	If not, why not?	Disinterest □; unpaid activity □; poor collaboration with foresters □; other reasons □ to be specified
72	What activities do you think people regularly carry out in the forest?	Livestock □; agriculture □; green wood cutting □; bush fires □; carbonization □; gold panning □; others to be specified □
73	Are there any wildfires in your area?	Yes □; no □; abstention □
74	If so, who lights the vegetation?	Hunters □; herders □; farmers □; foresters □; customary □; children □; heads of families □; others to be specified □
75	At what times of year do people set fire to vegetation	October-December □; January-February □; March-April □; other □ to be specified
76	Why do you light the vegetation fire at ?	Clearing fields □; keeping reptiles away from houses □; sacrificial rites □; hunting □; other reasons □ to be specified
77	How do you perceive the evolution of reserves?	Positive evolution □ ; regressive evolution □ ; stagnation □ ; no perception □
78	If the trend in reserves is downward, what do you think are the reasons?	Cutting green wood □; late bushfire □; charring □; gold panning □; grazing in the forest □; opening fields for agriculture □; insufficient rainfall; drought □; strong winds; high temperatures □; other □ to be specified
79	Do you know of any endangered plant species in the reserves?	Yes □; no □; abstention □
80	If so, which ones?	To quote
81	What do you propose to improve the conservation of biodiversity on the reserves?	To be specified
82	What do you think the French government is doing to protect forests?	Forest monitoring □ ; reforestation □ ; other □ to be specified
VI. Forest law enforcement and penalties		

83	Have you ever been a victim of non-compliance with forestry legislation	Yes □; no □; abstention □
84	If so, what did you really do?	Cutting green wood □; grazing in an area where activity is prohibited □; responsibility for a vegetation fire □; poaching □; other □ please specify
85	What penalties have you received?	Imprisonment □; fine □; other □ specify duration if imprisonment and sum if fine
86	If not, why not?	Intervention by a political or administrative authority in your favor □ ; Freedom in exchange for money, animals or materials □ Other □ please specify
87	What solutions do you propose to ensure that your business complies with forestry legislation?	To be specified

Appendix 2: Guides for semi-structured interviews and focus groups

Appendix 2a: Interview guide for transhumant herders

Sex: Male □; Female □

Age :

Marital status: Single □; Married □; Divorced □; Widowed

Profession:

Level of education: None □; literate □; primary □; secondary □; higher □

1. How long have you lived here?
2. What is the composition of your livestock?
3. What is the herd size by species?
4. Is there an area reserved for grazing?
5. If not, what do you do to feed the cattle?

6. Do you prune any plant species whose leaves are particularly prized by animals as food? If so, which ones?
7. What are the cutting zones?
8. Are there any cattle trails in your area? If not, what channels do you use to drive your herd to pasture? If so, are these tracks clearly and visibly marked? If they are, do you respect them?
9. Are you often confronted with difficulties related to the obstruction of cattle tracks by fields? If so, what do you do to reach the grazing area?
10. Are there any conflicts between you (breeders) and the farmers?
11. Do you often spend time in the forest with your herd during the dry season?
12. How do you perceive the evolution of reserves? What are the reasons for this?
13. Are there any extinct or endangered plant and animal species in the reserves?
14. Do you know that there is legislation governing logging in Burkina Faso? Is grazing authorized in the forest?
15. What activities are prohibited in the forest?
16. Have you ever been a victim of non-compliance with forestry legislation? What penalties have you suffered?
17. If you weren't sanctioned, what were the reasons?
18. What solutions do you propose to improve the way you do business, while complying with forestry legislation?

Appendix 2b: Interview guide for foresters

Sex: Male □; Female □

Profession:

Level of education: secondary □; higher □

1. Briefly tell us about the history of forestry legislation in Burkina Faso.

2. Can you tell us about the state of the Bontioli wildlife reserves?

3. What do you think of the state of deterioration of these reserves?

4. What are the reasons for deforestation in Bontioli's wildlife reserves?

5. What concrete actions are your agents taking in the field?

6. Do you issue logging permits to local people?

7. What conditions need to be met to qualify?

8. Are people aware of current forest legislation?

9. Are forestry regulations applied by local populations?

10. Do you initiate legal proceedings against those who infringe forestry regulations?

11. Are sanctions taken against them? What are the most common?

12. What mechanisms are in place to disseminate forestry legislation to local communities?

13. What do you do to ensure that forestry legislation is properly disseminated and understood by the local population?

14. What do you think of local people's application of forestry legislation?

15. What do you think of gold panning in these reserves? And what solutions would you recommend to remedy the situation?

16. What is your analysis of the overall use of the Bontioli wildlife reserves by local people?

17-How do people perceive these protected areas?

18. Why aren't these reserves developed for the benefit of the local population?

19. What do you suggest for an economic valuation of the reserves?

20- How can we enhance the cultural value of these reserves for the benefit of the local population?

21. What is the impact of poaching on wildlife in general and on pachyderms in particular in the Bontioli wildlife reserves?

21-Do you have any other experiences to share with us about the overall dynamics of these wildlife reserves?

Appendix 2c: Interview guide for gold miners

1. When did you start panning for gold on this site?
2. Where were you before you discovered this site?
3. Are there any "indicator elements" for gold in the forest?
4. Can you explain the gold extraction process?
5. How and with what products do you treat gold?
6. Is there a dedicated ore-processing area outside the forest?
7. Do you often process gold in rivers?
8. What do you do once all the ore has been extracted from the hole? Do you close it?
9. Given that the gold panning site is located in the forest, how do you reconcile this activity with maintaining the vegetation?
10. Have you reforested any areas that had already been logged?
11. Are you aware that the forestry code prohibits this activity in the forest?
12. What is your relationship with the environmental department?
13. Are you prepared to give up gold panning if the forest is developed?

Appendix 2d: Interview guide for hunting association managers

1. How long has your association been in existence?
2. What are its missions?
3. What exactly are your roles in the forest?
4. What time of year do you hunt in the forest?
5. And what are the means and/or hunting equipment?
6. What wild animals do you hunt?
7. How often do you use fires in your hunting activities?
8. What type of hunting (traditional, modern or mixed) do you do in the forest?
9. Are you familiar with forestry legislation in Burkina Faso?
10. How do you access the (total) reserves to carry out your hunting activities?

11. What is your association doing to protect the forest?

12. What are your perceptions of forest degradation?

13. In your opinion, are the reserves valued economically and culturally?

14. What are the impacts of forest degradation on the habitat and mobility of wild animals and pachyderms in particular?

15. What is the impact of poaching on the reserves' wildlife, especially pachyderms?

16. What solutions would you suggest to improve your activities on the reserves?

17. Do you have any other experiences you'd like to share with us about managing your forest?

Appendix 2e: Interview guide for the head of thc Cooperative Union

1. Cooperative name

2. Creation date

3. Receipt references (if available)

4. How many groups are there?

5 - What is the cooperative's membership (if possible)?

6. Your organization is represented in how many villages bordering the reserves?

7. What are the membership conditions

8. What strategies do you use to mobilize women?

9. Role and missions of the cooperative?

10. Which sectors are involved in your company's activities?

11. What activities do cooperative members carry out on the reserves?

12. How is the collection of non-timber forest products (NTFP) organized on reserves?

13. What fund-raising strategies have you put in place to ensure the smooth running of your organization?

14. How are the funds generated by the cooperative's activities distributed?

15. How do the members of the structure contribute to protecting and restoring the reserves?

16. What are the prospects in terms of organization for the structure under your responsibility?

17. Do you have any other experiences you'd like to share with us?

Appendix 2f: Focus group guide for sorghum beer (dolo) producers

1. When do you make sorghum beer

2. What type of fuel do you use to make dolo?

3. What kind of fireplaces do you use for this activity?

4. How do you obtain the wood? Do you collect it from the bush or do you buy it?

5. Tell us about the availability of firewood in your village to carry out this activity.

6. How many bundles of wood do you use for a sorghum beer-making session?

7. Where does the wood come from? Bush (old fallow land)? Fields? Or forest?

8. Are you authorized to take timber from the forest? What are the conditions?

9. Are there people authorized to cut and sell wood?

10. Are there any places where you can buy wood?

11. Is wood affordable? How much does a bundle of wood cost?

12. Are you aware of the existence of forestry legislation?

13-How do you assess the application of forestry legislation by the general public and sorghum beer producers in particular?

14-What solutions would you suggest to reduce the amount of wood cut by your business?

Appendix 2g: Focus group guide for NTFP farmers and processors

1) What NTFPs do you harvest?

2. At what times of year do you use these products?
3. What equipment is used to harvest these products?
4. Do you often pick before ripening?
5. How do you collect produce (fruit, seeds or leaves) from branches far from the ground?
6. What are the main products you process?
7. Can you briefly explain their transformation process?
8. Where do you collect the wood for cooking these products?
9. Are you authorized to cut wood in the forest to carry out your activities? What are the conditions?
10. How much wood (in bundles) do you use to process these products (depending on the quantity to be leathered)?
11. Did you know that cutting green wood in the forest is regulated by the forestry code?
12. Have you ever benefited from awareness-raising campaigns on the conditions under which NTFPs and wood are collected in the forest?

14. What is your relationship with Eaux et Forêts agents?

13. What solutions do you propose for the healthy harvesting and processing of NTFPs in compliance with forest legislation?

Appendix 3: Floristic survey form

Observer name: .... Survey date: ....

Survey number: ....Plot coordinates: ....

Plot size: .... Survey time: ....mn

Plant formation: FG □; SB □; SA □; Sa □; PA/J □

N°	Botanical name	CHP (cm)	Ht (m)	Tree health

FG: gallery forest; SB: wooded savannah; SA: wooded savannah; Sa: shrubby savannah; PA/J: agroforestry park/fallow. 1: ligneux saint; 2: ligneux ébranché; 3: ligneux brûlé; 4: presence of holes in the wood; 5: ligneux semi-dry; 6: ligneux mort sur pied

Appendix 4: Observation form

Slash	Visiting period	Comments

Appendix 5: Projected resident population

Villages	2010	2011	2012	2013	2014	2015	2016	2017	2018	2019	2020
Commune of Tiankoura (Bougouriba province)											
Diourao	620	639	659	680	701	722	744	766	789	812	836
Kanséo	330	340	351	362	373	384	396	408	420	432	445
Kimpéo	142	146	151	155	160	165	170	175	180	186	191
Kourguenou	152	673	694	715	737	759	783	806	830	855	880
Sinkiro	232	239	247	254	262	270	278	287	295	304	313
Tiopanao	294	304	313	323	333	343	353	364	375	386	397
Diébougou commune (Bougouriba province)											
Mouvielo	2046	2110	2176	2243	2312	2382	2455	2528	2604	2681	2760
Nako commune (Poni province)											
Babière	305	314	324	334	344	354	365	376	387	398	410
Barro	383	395	407	419	432	445	458	472	486	500	514
Boukairo	812	837	863	889	916	944	972	1001	1030	1061	1092

Gourkpè-soum	553	570	588	605	624	643	662	681	701	722	743
Hemkoa	167	172	177	183	188	194	200	206	212	218	224
Tiékar	479	493	508	524	540	556	573	590	607	625	643
Commune of Dissin (Ioba province)											
Done	1190	1221	1254	1287	1322	1357	1393	1430	1468	1506	1546
Villages	**2010**	**2011**	**2012**	**2013**	**2014**	**2015**	**2016**	**2017**	**2018**	**2019**	**2020**
Commune of Zambo (Ioba province)											
Bobra-Gogo	865	888	912	936	961	986	1012	1039	1067	1095	1124
Bolkpe	320	321	338	347	356	365	375	385	395	406	416
Bontioli	1270	1304	1339	1375	1411	1449	1487	1527	1567	1609	1652
Dergane	288	295	303	311	319	328	337	346	355	364	374
Djikologo	1616	1659	1706	1749	1795	1843	1892	1942	1994	2046	2101
Forotéon	1314	1349	1385	1422	1460	1499	1539	1579	1621	1664	1709
Gamgbo	519	533	547	562	577	592	608	624	641	658	675
Gbongbo-	881	905	929	954	979	1005	1032	1059	1087	1116	1146

zodoun											
Gbonko	948	973	999	1026	1053	1081	1110	1139	1169	1200	1232
Habre	1321	1356	1392	1429	1467	1506	1546	1587	1629	1673	1717
Korompéri	743	763	784	805	826	848	870	894	917	942	967
Manoa	1293	1328	1363	1400	1437	1475	1514	1554	1596	1638	1682
Nambléteon	388	398	409	420	431	443	454	466	479	492	505
Pintou	385	395	406	416	428	449	451	463	475	487	500
Tieretéon	171	175	180	185	189	195	200	205	210	216	222
Tovor	1255	1289	1323	1358	1394	1431	1469	1508	1548	1590	1632
Zoolo	1602	1644	1688	1733	1779	1827	1875	1925	1976	2028	2082
Total	23384	24036	24712	25401	26106	26630	27573	28332	29110	29910	30730

Source: INSD, 2009

Appendix 6: Transition matrices of land-use unit areas in ha

Table: RTFB land-use transition matrix between 1986 and 2002

Thematic classes	Rainfed cultivation	Gallery forest	Wooded savannah	Shrubby savannah	Bare ground	Total 1986
Rainfed cultivation	354,54	22,63	3158,87	4670,77	0,27	8207,09
Gallery forest	0,00	565,65	0,00	0,00	0,00	565,65
Wooded savannah	32,99	0,00	326,33	149,20	0,00	508,52
Shrubby savannah	131,50	0,05	1827,38	2220,24	0,00	4179,18
Bare ground	0,00	0,00	0,27	0,09	0,00	0,36
Total 2002	519,04	588,33	5312,85	7040,30	0,27	13460,80

Source: Landsat images from 1986 and 2002

Table: RPFB land-use transition matrix between 1986 and 2002

Thematic classes	Fields and fallow land	Gallery forest	Wooded savannah	Shrubby savannah	Bare ground	Total 1986
Fields and fallow land	3837,62	21,17	4986,72	11618,58	1,02	20465,11
Gallery forest	0,58	1554,06	12,01	9,90	0,00	1576,55
Wooded savannah	118,59	0,00	256,54	167,04	0,00	542,17
Shrubby savannah	1333,44	8,06	3090,17	5162,37	0,33	9594,37
Bare ground	0,43	0,00	0,17	0,21	0,00	0,81
Total 2002	5290,66	1583,29	8345,61	16958,10	1,34	32179,01

Source: Landsat images from 1986 and 2002

Table: RTFB land use transition matrix between 2002 and 2018

Thematic classes	Fields and fallow land	Gallery forest	Wooded savannah	Shrubby savannah	Bare ground	Total 2002
Fields and fallow land	4,66	6,23	69,70	400,98	36,94	518,51
Drill gallery	0,79	0,00	31,76	509,06	45,24	586,86
Wooded savannah	21,29	219,07	1351,25	3531,44	179,42	5302,46
Shrubby savannah	22,28	79,03	865,06	5615,81	447,22	7029,40
Bare ground	0,00	0,00	0,00	0,18	0,09	0,27
Total 2018	49,02	304,32	2317,77	10057,47	708,91	13437,50

Source: Landsat images from 2002 and 2018

Table: RPFB land-use transition matrix between 2002 and 2018

Thematic classes	Fields and fallow land	Gallery forest	Habitat	Wooded savannah	Shrubby savannah	Bare ground	Total 2002
Fields and fallow land	1252,07	91,18	123,01	129,39	2271,08	1407,70	5274,43
Gallery forest	78,33	33,17	8,23	86,09	950,70	420,47	1577,00
Wooded savannah	837,26	628,10	141,43	785,21	4569,09	1353,29	8314,39
Shrubby savannah	1296,50	444,39	251,95	531,25	9216,32	5158,84	16899,26
Bare ground	0,39	0,24	0,00	0,09	0,36	0,27	1,34
Total 2018	3464,55	1197,08	524,62	1532,04	17007,54	8340,58	32066,42

Source: Landsat images from 2002 and 2018

Table: RTFB land use transition matrix between 1986 and 2018

Thematic classes	Fields and fallow land	Gallery forest	Wooded savannah	Shrubby savannah	Bare ground	Total 1986
Fields and fallow land	33,63	209,69	1166,87	6281,16	502,45	8193,79
Gallery forest	0,79	0,00	31,76	490,04	41,94	564,53
Wooded savannah	0,48	9,20	114,79	376,50	6,47	507,44
Shrubby savannah	14,12	85,43	1004,09	2909,69	158,05	4171,38
Bare ground	0,00	0,00	0,27	0,09	0,00	0,36
Total 2018	49,02	304,32	2317,77	10057,47	708,91	13437,50

Source: Landsat images from 1986 and 2018

Table: RPFB land-use transition matrix between 1986 and 2018

Thematic classes	Fields and fallow land	Gallery forest	Habitat	Wooded savannah	Shrubby savannah	Bare ground	Total 1986
Fields and fallow land	2515,93	659,21	415,66	745,77	10247,57	5810,57	20394,73
Gallery forest	73,08	33,17	8,23	86,09	963,51	406,17	1570,25
Wooded savannah	33,53	40,46	2,12	44,32	356,93	63,13	540,50
Shrubby savannah	842,01	464,18	98,61	655,84	5438,96	2060,53	9560,14
Bare ground	0,00	0,05	0,00	0,00	0,57	0,18	0,81
Total 2018	3464,55	1197,08	524,62	1532,04	17007,54	8340,58	32066,42

Source: Landsat images from 1986 and 2018

Appendix 7: Lists of plant species inventoried in the two reserves

Appendix 7a: List of plant species inventoried in RTFB

Order no.	Botanical name of plant species
1	*Acacia dudgeoni* Craib ex Hall.
2	*Acacia laeta* R. Br. Ex Benth
3	*Acacia macrostachya* Reichenb. ex DC.
4	*Acacia sieberiana* DC.
5	*Adansonia digitata* L.
6	*Afzelia africana* Smith ex Pers.
7	*Anogeissus leiocarpus* (DC.) Guill. et Perr.
8	*Balanites aegyptiaca* (L.) Del.
9	*Berlinia grandiflora* (Vahl) Hutch. et Dalz.
10	*Bombax costatum* Pellegr. Et Vuillet
11	*Burkea africana* Hook. f.
12	*Cassia sieberiana* DC.
13	*Combretum collinum* Fresen.
14	*Combretum fragrans* F. Hoffm
15	*Combretum glutinosum* Perr. ex DC.
16	*Combretum micranthum* G. Don
17	*Combretum molle* R. Br. ex G. Don
18	*Combretum nigricans* Lepr. ex Guill. et Perr.
19	*Crossopteryx febrifuga* (Afzel. ex G. Don) Benth.
20	*Daniellia oliveri* (Rolfe) Hutch. et Dalz.
21	*Detarium microcarpum* Guill. et Perr.
22	*Diospyros mespiliformis* Hochst. ex A. Rich.
23	*Ekebergia senegalensis* A. Juss.
24	*Entada africana* Guill. et Perr.

25	*Gardenia erubescens* Stapf et Hutch.
26	*Gardenia ternifolia* Schumach. et Thonn.
27	*Hexalobus monopetalus* (A. Rich.) Engl. et Diels
28	*Khaya senegalensis* (Desr.) A. Juss.
29	*Lannea acida* A. Rich.
30	*Lannea microcarpa* Engl. et K. Krause
31	*Lannea velutina* A. Rich.
32	*Mitragyna inermis* (Willd.) Kuntze
33	*Parkia biglobosa* (Jacq.) R. Br. ex G. Don
34	*Pericopsis laxiflora* (Benth.) van Meeuwen
35	*Piliostigma thonningii* (Schumach.) Milne-Redh.
36	*Pseudocedrela kotschyi* (Schweinf.) Harms
37	*Pterocarpus erinaceus* Poir.
38	*Saba senegalensis* (A. DC.) Pichon
39	*Sarcocephalus latifolius* (Smith) Bruce
40	*Sterculia setigera* Del.
41	*Strychnos innocua* Del.
42	*Strychnos spinosa* Lam.
43	*Tamarindus indica* L.
44	*Terminalia avicennioides* Guill. et Perr.
45	*Terminalia macroptera* Guill. et Perr.
46	*Trema orientalis* (L.) Blume
47	*Vitellaria paradoxa* Gaertn. f.
48	*Vitex doniana* Sweet
49	*Ximenia americana* L.

Appendix 7b: List of plant species inventoried in the RPFB

Order no.	Botanical name of plant species
1	*Acacia dudgeoni* Craib ex Hall.
2	*Acacia ehrenbergiana* Hayne
3	*Acacia gourmaensis* A. Chev.
4	*Acacia laeta* R. Br. Ex Benth.
5	*Acacia macrostachya* Reichenb. ex DC.
6	*Acacia polyacantha* Willd. subsp.
7	*Acacia senegal* (L.) Willd.
8	*Acacia seyal* Del.
9	*Acacia sieberiana* DC.
10	*Adansonia digitata* L.
11	*Afzelia africana* Smith ex Pers.
12	*Anogeissus leiocarpus* (DC.) Guill. et Perr.
13	*Balanites aegyptiaca* (L.) Del.
14	*Berlinia grandiflora* (Vahl) Hutch. et Dalz.
15	*Bombax costatum* Pellegr. Et Vuillet
16	*Combretum collinum* Fresen.
17	*Combretum fragrans* F. Hoffm.
18	*Combretum glutinosum* Perr. ex DC.
19	*Combretum molle* R. Br. ex G. Don
20	*Combretum nigricans* Lepr. ex Guill. et Perr.
21	*Cordia myxa* L.
22	*Crossopteryx febrifuga* (Afzel. ex G. Don) Benth.
23	*Daniellia oliveri* (Rolfe) Hutch. et Dalz.
24	*Detarium microcarpum* Guill. et Perr.
25	*Diospyros mespiliformis* Hochst. ex A. Rich.
26	*Ekebergia senegalensis* A. Juss.

27	*Entada africana* Guill. et Perr.
28	*Faidherbia albida* (Del.) Chev.
29	*Ficus on* Forssk.
30	*Ficus sycomorus* subsp.
31	*Gardenia erubescens* Stapf et Hutch.
32	*Gymnosporia senegalensis*
33	*Hexalobus monopetalus* (A. Rich.) Engl. et Diels
34	*Khaya senegalensis* (Desr.) A. Juss.
35	*Lannea acida* A. Rich.
36	*Lannea microcarpa* Engl. et K. Krause
37	*Lannea velutina* A. Rich.
38	*Lonchocarpus laxiflorus* Guill. et Perr.
39	*Manilkara multinervis* (Bak.) Dubard
40	*Mitragyna inermis* (Willd.) Kuntze
41	*Parkia biglobosa* (Jacq.) R. Br. ex G. Don
42	*Pericopsis laxiflora* (Benth.) van Meeuwen
43	*Piliostigma reticulatum* (DC.) Hochst.
44	*Piliostigma thonningii* (Schumach.) Milne-Redh.
45	*Prosopis africana* (Guill. et Perr.) Taub.
46	*Pseudocedrela kotschyi* (Schweinf.) Harms
47	*Pterocarpus erinaceus* Poir.
48	*Saba senegalensis* (A. DC.) Pichon
49	*Sarcocephalus latifolius* (Smith) Bruce
50	*Sterculia setigera* Del.
51	*Strychnos spinosa* Lam.
52	*Tamarindus indica* L.
53	*Tectona grandis* L. f.
54	*Terminalia avicennioides* Guill. et Perr.
55	*Terminalia macroptera* Guill. et Perr.

56	*Trichilia emetica* Vahl
57	*Vitellaria paradoxa* Gaertn. f.
58	*Vitex doniana* Sweet
59	*Ximenia americana* L.

Appendix 8: Speculation by riverside residents

Speculations produced	Frequency (%)
Corn	16,20
Small millet	15,70
Red sorghum	15,40
Peanut	11,60
Cotton	10,50
Bean	8,20
Sorrel	4,70
White sorghum	4,60
Squash	4,30
Tomato	1,80
Chili pepper	1,20
Sesame	1,10
Onion	1
Cabbage	1
Cowpeas	0,90
Lettuce	0,50
Peppers	0,30
Other	0,80

TABLE OF CONTENTS

MIX
Papier aus verantwortungsvollen Quellen
Paper from responsible sources
FSC® C105338

Printed by Books on Demand GmbH, Norderstedt / Germany